"NEXUS"

Note :

#Este libro narra una historia de amor, pero no una historia de amor cualquiera…es un amor que es capaz de cambiar el mundo,el amor de un padre por su hija que traspasa todo espacio,tiempo,frontera,nacion,raza y universo…hehe!!!.

This will be as "Dune " was..well.much better…hehe!!!.Its the beginning if a saga and the second one "Xenus" will be the genesis of this saga,as the second one Dune (sons of Dune was of..),and it can have a very similar structure…though …I keep on with "Los Niños Perdidos" structure but (scheme) but I can use these "prologues" at every chapter or Episode…huahuah!!! Or "Legend"…hauhauhua!!! And

"Xenoarqueoastronomy" is the original human resources to travel anywhere in the cosmos without moving.Human race has the "Hardware" in our Soul-Matrix to do this and discover all the universe,we don´t need machines to do this.In this investigation we found a main program called "NEXUS".

 "All Alien Races for Hundred hundred Thousand years have come here, at Earth Planet infiltrating shape-shifted and hybrids human beings..and this goes on… Thats the reason there are no sources from aliens as aliens in all our history and that's the point where my story beggings,my Exocosmobiological theory ". We cant see them "originally" and when It happens they are close their ships,but,all of them,intraterrestrials,Vlash aliens,Akart do shapesifting to humans bodies (dress up) and Hybrids (genetical make up).

This book is dedicated to my Cosmic twin soul "Patricia Tanambi Kunha"…hehe!!!.With all the love of the Universe…hehe!!!.

These are our Eridani´s human extraterrestrials brothers and sisters at Eridani´s Resistance Central Command:

1-Metah-Tron-Team= "Kate´s Hair "
Friends:
-Tatdi 314
-Vehu-I-Ah
-Hel-Eal
-Sit-Hael
-El-Emiah
-Meiha-Siah
-Lilael
-Ei-Chaiah
-Keah-Et-Hel (She).

2- Rat-Sel-Team- "Ho-Out-Mat"- "The Kae-Robins friends":

-Hiclel
-El-Dah
-Lao-Veah
-Ha-Eah
-Yesal
-Embi-Hel
-Harel
-Hek-Meh

3-Tex-Fel-Team="Tron-Es"
Friends:
-Alub-Ha
-Kalel
-Lub-Ha
-Phle-Ah
-Em-El-Tixel
-Eaal
-Em-El-Hal
-Ahi-U-Ah

4-Tsa-De-Kel-
Team="Dominion"
Friends:
-Neta-Hah
-Ah-Iah
-Earth-L
-She-Ah
-Rul
-Omal
-El-Kabla
-Ba-Sheah

5-Kameal-"Team :"The Mighty-Ones":

-Ea-Hueah
- Lê-He-Ah
- Xá-Bakah
-Man-Del
-A-Nel
-Ah-Meah
-Rahal
-Ea-Ecli

6-Ra-Fal- Team "Tefer-Te":"The Virtus Est Ones"
Friends:
-Ah-Nael
- Makel
- Beul-Aah
-U-Elah
-Selenah
-Rel
-Saleah
-Meahal

7-Ah-Nel- Team "Net-Ci-Di"":"The Principle Ones":
Friends:

-Behul
-Da-Enel
-A-Kah
- Emah-Malah
- Na-Nal
- Nat-Hael
- Mebaha
- Poel

8-Mak-El- Team "He-Od":"The Rangles":

-Nma-Meah
- Eal
-Ah-Ralme
- Me-Trel
-Umab-El
- Ea-Hehel
- Enul
- Mehel

9-Ual-Bral-Team "Ues-Da":"The Noodle Ones":

-Dim-Beah
- Man-Kla
-Aul
-Ahbu-Heah
- Roxlua-Bemeah
-Uab-Meha
-Ahi-El
-Mum-Eah.
- Emah-Hel...hehe!!!.

-This is a massive Attack!.
-With my mantis Queen,i´ve broken all"religion" rules…huahuhaua!!!.The Living God has
taken me !!! And has told me : Don´t be Silly!!! You are totally free!!!,I create you totally free,I
discover my Own into You!!!.So, If you Are tied..I only ask U: Remember
me!!!...Huahuahua!!!.
TO BE CONTINUED…Troy, Troy, and Troy again The events of ancient Troy connect
fundamentally with the death of Princess Diana and so much else besides. The Merovingian
bloodline goes back to the Trojan War and beyond, and it was they who founded the city of
Paris and named it after the Trojan called Prince Paris, the lover of Helen (El-en) of Troy.
According to Barbara Walker in The Woman's Encyclopedia Of Myths And Secrets, Helen was
said to be an incarnation of the virgin Moon-goddess and daughter of Hecuba or...Hecate.20
Helen was also known as Helle and Selene and was worshipped at a Spartan sexual festival
called the Helenphoria that included sexual symbols carried in a basket called the helene.21
Troy or Troia means "three places" in Greek and Hebrew and almost certainly relates to the
triple-goddess symbolism of Atlantis and Lemuria with one deity divided into the "trinity", or
three aspects. Hecate was known as "Hecate of the Three Ways".22 Troy or Troia is also the
origin of the name Tripoli, the capital of Libya, which is so associated with the reptilian
Amazons. The legends of Troy say that Helen married the "Moon-king" Menelaus, who was

promised immortality because of this "sacred marriage". When Helen went off with Prince Paris, Menelaus wanted to protect his immortality and the wealth the marriage had secured, and sailed with his army to take her back. This was the war between the male-led Greeks and the goddess-led Trojans. Many high Satanic priestesses take the name Helen, Helena, or Elaine - El-aine.* It was under the name Elaine or Ellen that Helen of Troy became the symbolic queen of Britain in pagan times. As I outlined earlier, it was a relative of Helen, a Trojan called Brutus, who sailed west to Britain after the fall of Troy and founded a city called Caer Troia or New Troy - today's London. Other derivatives of El or Hel are Helenia, Helga, Hild, Helsinki, Holstein, and Holland (Hel-land or Halland), one of the major centres for the reptilian bloodlines to this day. Pliny, the Roman writer, said that all the people of "Scatinavia", or Scandinavia, were children of "Mother Hel" and were called Helleviones.24 They believed that she lived in elder or Hel-trees/elven trees. Sir Laurence Gardner, of the Royal Dragon Court and Order, says that his "Dragon Bloodlines" have been called the Elven Race and that terms like elf, fairy, and pixie all symbolise the "representatives of various castes within the kingly succession" (the reptilian hierarchy).24 So many fairy tales and other children's stories are encoded with the theme of the dragon bloodlines and their battles for power. The tales of princes and princesses "turning into a frog" is symbolic of shape-shifting. The same with dragon princesses locked in towers or giving birth to frogs. (FROM DAVID ICKE...CHILDRENS OF THE MATRIX...HEHE!!!)
*

"I send you this message through Patricia because they have let me without Whatssapp and nothing...Here i´ve get some money to the cell phone to have Internet 24mech.hours but I need to make a plan more complete,so we must wait til I found the solution,a little bit...hehe!!!...And we´ll do everything til now because of these problems...hehe!!!.Mummi,I miss you so much with this love that ties us both also with Patricia in this Cosmic Mission...HEHE!!!".
-This message is so important,father?
-It changed the course of all Universe.
-Really?.How?.
-Do you want to know? Here´s the Story...hehe! !:
"This is a Tale about Espylon Eridani.":
In an ancestral time, when warriors and witchcrafts were in the land of Eridani, there was a legend....And you could ask...about what?..About Kings and Thieves,Lords and fellows...hehe!!!.
Epsylon Eridani:

The Eridanians have taking the most action to help their brothers and sisters here on Earth ,in essence interfering with the saurian "interferers" from the Draconis ,Bootes,Reticuli,Canis,etc..Constelations and the "Solar Tribunal" groups of Mars,Moon ,Saturn,etc.and so on.

Many of the Reptiloids have shown animosity towards humans, . One "device" had a small orange light on it. When this is touched, they simply "disappear" (if they were draining energy to manifest, like happens in various kinds of apparitions The head is similar to the Reptilian Humanoids on the TV Program "BABYLON 5". In fact, the REPTOIDS are built like Big Foot #PHISICAL CHARATERISTICS OF REPTOIDSThe scales were smaller where the body is flexible, like around the elbow. They are "Telepathic". Very quick thoughts. #YHE WAY OF THEIR TELEPHATHY

IDEAS #SYMBOLIC TELEPHATHIC LANGUAGEdriven by Images and geometric Impressions. NOT a "linear" form of communication, like words. More, of a Symbol or Image Language. They DO react to your "thoughts". They can "overwhelm" you with DATA (it seems like "thought compression"). It can take you a long time to assimilate their "TRANSMISSION" and integrate it. After awhile I "saw" the language, as FRACTAL "Mandelic" Hyper-Spheres or "THOUGHT FORM" constructs, to pass But, if you Fear them, they will taunt your weakness and will DISrespect you. Careful, don't DIS-"REP" them either. They are no better than you. I "glean" they want something WE haveDATA. These sucks are data freaks. Highly Intellectual. Limited Emotional response (Hate, Fear and a "short circuit") Confusion state (mild Panic). You can think or do "the UNEXPECTED" and they do "freak out". So...stay calm and centered. They RESPECT that. Then they "THINK" you are dangerous. #PERO SI LES TIENES MIEDO VAN A USAR TU DEBILIDAD Y TE HUMILLARAN..PERO EN CASO CONTRARIO NUNCA LOS DES-RESPETES...HEHE!!! #ELLOS NO SON MEJORES QUE TU#ELLOS QUIEREN ALGO QUE TENEMOS...ALGO EN LA NATURALEZA DE NUESTRA ALMA...PODEMOS VOLVER A LAS ONDAS ZETA DE EQUILIBRIO Y VOLVER A A LA FUENTE..ELLOS NO ANALIZAN,SON DESEQUILIBRADOS..LES GUSTA COCAINA,OPIO,HONGOS MAGICOS Y LIQUIDOS HUMANOS PARA SUS ENDORFINAS...NO PUEDEN ENTRAR EN TU MENTE SI TU ANTES NO HAS ENTRADO EN LA SUYA...CONOCEN MUCHO ACERCA DE LAS FORMAS EXTERNAS DE LA CREACION...CONOCEN LA FUERZA...PERO DETRAS DE ELLA ESTA LA FUENTE...NOS NECESITAN,PERO NOSOTROS NO LES NECESITAMOS A ELLOS....TEMEN LA VUELTA DEL DRACO CORNUDO Y CON ALAS..EL DRACO MAS IMPORTANTE ES ALBINOabduction in which he and his grandmother were taken to a spacecraft in the company of reptilian alien the majority of the "grey" type alien entities actually possess a REPTIL/SAURIAN genetic base) Brain capacity is 1800 cc versus 1300 cc for the average human. The skin is grey or ashen and under the microscope appears meshlike. This meshlike appearance gives it the reptilian texture of granular skinned lizards like iguana or chameleon. There was a colorless liquid in the body without red cells, no lymphocytes, no hemoglobin. There was no digestive system, intestinal, alimentary canal, or rectal area in the ET autopsy.

Go Back   Brad Stieger he reason why the most frequently reported UFOnauts resemble REPTILIAN or AMPHIBIAN humanoids may be because that is exactly what they are, s. A provocative theory is that the dinosaurs didn't really vanish, they 'evolved' into a humanoid creature that eventually ran it's course, or was destroyed in an Atlantis-type catastrophe. (Such as the Great Deluge - Although Steiger and others may hold to an 'evolutionary' hypothesis, this may not necessarily be the case, especially when the 2nd Law of Thermodynamics and the laws of entropy are brought to bear. Instead of 'evolving' from a far less complex form, it is in fact far more likely that the serpent race MUTATED via atrophication, natural selection, environmental adaptation, survival of the fittest and most intelligent, and possibly a bit of superficial molecular shape-shifting IF as some believe regressive supernatural entities have been involved in guiding the 'evolution' of the 'serpent race' down through history... into it's various known and unknown branches, from a complex bi-pedal single species which originally inhabited the earth in ancient times#INTERESANTE TEORIA DE BARNTON QUE SEÑALA QUE LOS DINOSAURIOS NO DESAPARECIERON SINO QUE "EVOLUCIONARON"EN UNA FORMA HUMANOIDE..CON LA AYUDA DE ENTITADES SBOBRENATURALES,REGRESIVAS QUE HAN ESTADO OCUPADAS-PREOCUPADAS-RELACIONADAS EN GUIAR LA EVOLUCION DE LA RAZA DE LA SERPIENTE EN SUS RAMAS GENETICAS CONOCIDAS Y NO CONOCIDAS A LO LARGO DE LA HISTORIA...Y QUE HABITABAN LA TIERRA EN TIEMPOS ANTIGUOS Y DESAPARECIERON EN UNA CATASTROFE TIPO LA ATLANTIDA...Y QUE POSOBLEMENTE FORMEN EL GRUESO DE LOS OVNINAUTAS QUE MUCHOS TESTIGOS HAN VISTO DURANTE LOS ULTIMOS 50 AÑOS...HEHE!!!(ESO ES MIO..PERO BUENO LO MEZCLO CON EL ESTILO DE BRANTON Y QUEDA PERFECTO...HEHE!!!)
#NOTAS :
#

07/08/2015 CAL GREG:
-1)HUMAN SACRIFICES …AZTEC TIMES..2001…WE ARE NOT AT THE PEAK OF THE FOOD PYRAMID..THE FOOD CHAINS…..TO PUT AN ALIEN AT HRE PEAK..HH AND A HUMAN TALLER THAN THESE ALIENS…HH
2) EL MESSENGER SALVÓ AL MUNDO (07/08/2015 CAL GREG)
3)I CAN KNOW IF IS OR NOT…IF READS OR NO MY MESSAGES…HH
4)LA TRADICION DE LOS MAGAMILEDONICOS…HH
5)26:04 -32:22MIN, MEC. AHK PART 25 THE BEST FOR THE LAST…HEHE!!.
-6)**********THIS IS THE MOST IMPORTANT CHANGE IN ALL MY WHOLE LIFE…HEHE!!!...NOTA DEL 07/08/2105 CAL GREG:
#YA ESTÁ CLARO!!!!:
-we were created originals as humans and totally free!!!!
-Los Creadores de los S.E.R.ES. (LITERAL) HUMANOS SEGÚN EL LIBRO ARB SON LOS KURS (QUE HAN VUELTO AHORA Y QUE SE RELACIONARON SOBRE TODO CON LOS SUMERIOS-ABRAHAM) Y QUE SON LOS INSEKTOS!!!!...Y FUERON LOS CREADORES DE LOS SERES HUMANOS ,LOS FUNDADORES DE LA RAZA HUMANA,SEGÚN MIS INVESTIGACIONES Y CON LOS QUE ESTOY COMUNICADO DESDE 1999 TRAS MI PRIMERA CONEXIÓN CON LOS KURS EN EL DESIERTO DE OURZAZAT(MARRUECOS) BASE JEDI KURS….LA CLAVE LA HA ENCONTRADO DANTE SANTORI EN LA PELICULA "EL JUEGO DE ENDER 2013) CUANDO APARECE LA REINA INSEKTO IGUALÇ QUE LA FOTO DE LOS KURS DEL LIBRO ARB …HH 36:52 M.MEC. VIDEO 25 AUERDATE DE LA PALABRAS "ESTE ES TU FIN!" FRASE VLASH Y TODAS LAS ESTRELLAS DEL CIELO CAYERON Y APARECIO ESA CARA DE ESTRELLAS CON ESA FRASE Y LUEGO ENSEGUIDA APARECIÓ MI "ALIADO" KURS-INSEKTO YA PARA SIEMPRE ..9 MESES DESPUÉS NACIÓ SOPHIA COMO

RESULTADO DE OPERACIONES GENÉTICAS ALIEN-KURS HIBRIDASSUMAMENTE PREPARADAS CON ANTELACION POR LA GUERRA ENTRE LOS KURS Y LOS VLASH Y QUE TIENE COIMO ESCENARIO NUESTRO PLANETA Y LA RAZA HUMANA TAMBIEN….HEHE!!! COMO HIBRIDOS DE AMBAS RAZAS CON ELLOS…HEHE!!!

..and who was xenus?...the genesis of the human race, he/she could totally original from a conception came from insekts but form exoskeleton toendoeskeleton they realized that is better to long into time has this sturcture,flexible to accommodate this being better to different environmental places and circumstances…human are not flesh bodyt but soul-matrix and from inside to ourside it becomes human…and it realizes this being was different …quite different,he could heal himself from inside..and change everything with only his mind or soul…and human soul is only a form…a form from the source…from the essence…human soul is not the essebce but has the essence..and since the beginning his free will take it to grow it or squezze it!!!.And this circunstane was a big proff to our enemies,…who dare to mess with our DNA trying to find the essence and destroy it..this has been their main goal….and find the source and destroy it!!!...thets their mission and agendas…they don't have the conection with thr source though they had it in the beginning but they choose to forget it…they create a new creation..an artificial creation…but well talk about all this at second volume.."xenus".

In this well talk about robots primordial robots who were made from the original exoskeleton´s idea..but they failed..NEphilins..or they fallen down NEPHAL..to fall at Hebraic.since then,this organical robots…as their insekt creators we foung the humans..in all mythologies is said…"corn men" but before "wood men" were created…and at popol vuh is said…and is Nephilins..later on humans were a second creation…and it worked!!!.

10)LA UNION ENTRE LOS INSEKTOS-KURS Y LOS HUMANOS DE EPSYLON LO ENCONTRAMOS EN LA TIKRAZIA INSEKTO,LA GEOLOGIA NEURAL Y EL SIGLO XXIX (TRADUCIRLO AL INGLES) .
11) EPSYLON UNIDOS CON LOS ARCTURIANOS Y LOS MAYA GALÁCTICOS….HEHE!! EN RELACIONA ESTE SISTEMA KURS-VLASH.
12)VLASH ASHAMEL DEMONS..INSEKT HUMANS EPSYLON ANGELS.
13)QUIENES SON LOS INSEKTOS KURS Y LOS ANNUNAKI?
14)LOS INSEKTOS Y JESUS O DHOR KRYSTIL.,..HEHE!!!>
-15) 1:11:26 LA FOTO THE FLY INSEKTO THE KURS…HH
16)1:13:23 M.MEC.THE FLY INSEKTO-MECHANICAL MACHINE THE LAST FUSSION…HH.
17) Porque lo llaman espiritualidad cuando es Metalurgia ?,…hehe!!!>
18)1:32 tienda de discos y 0:35 -0:37 -0::42 m.mec y 0:44 m mec.
19) 06:15 m mec foto de la película "Zen".
20) LOS OJOS DE LOS HIBRIDOS DE LOS VLASH BRILLAN EN LA OSCURIDAD COMO LOS DE LAURA LA MADRE DE MIS HIJOS EN LA DISCOTECA DEL SALER EL AÑO 1999,CREYENDO QUE ERA SIRIANA-A ERA VLASH O SEA SIRIANA-B…HH.
21) WHATS SDA ¿ SEGÚN LA TERMINOLOGIA DE DANTE?...HEHE!!!.
22)"My brother,Alejandro,were teletransported at many times,in just one occasion he arrived 300 miles away home,CASTELLON,..andi many times close to his dog (Belgium Shepherd) called "Bicha"(snake at Andalucia-Spain)…
23) Eve was part of our mothership or my mothership and when ..i don't know what happened she disappeared or any one of us were brought all around the cosmos,..i don't know, I only know I lossed Eve for ever…or I thought this,and find her now is my biggest happiness ,but a planetary mission,that's because everytime I meet her im so happy!!!....hauhauhuahua!!!.
24) Writing this book as a tale or tale format…hehe!!!.
25) I know I unveiled Vlash Agenda to control all SDA in this war ,and they didn't expect I could remember everything and my reaction…hehe!!!.
26) I must progrees where I leaved it, so, with my own accounts,and with no ARB needing,but as testing them before,that's the way I must write the Human History from that Thesis, with my own accounts (Modular Center Epsylon Galaxy) (and what would happen if I take some item

from ARB?—YOU CAN DO IT!).but you must carry on where you stop it,so,with my own intuition,.At NEXUS---XENUS you MUST go on with your own resources (EVERYBODY KNOWS EVERYTHING but mass media and our consensual life covers our own normal knowledges discoverings) but the truth is everybody know everything ,and you can take all Dante Santori ´s accounts but not to fill spaes or verify beause my work is giving new knowkedge or open new spaes to new knowledges (unknown) (as clouds-ships) and anything else,as Dhor Krystil,Magdalena´s mary,mudras ,black swan,cross,X,wolf,Hybrids…but to link what I know from Reptilians,Greys,and their technologies,and Dante Santori don't take them,!,underground bases accounts,the NOW,3 Alternative,4..ALL THIS (by one side underground bases informations and to the other side Danta Santori accounts) but I don't give a damm to them in comparison what I know!!!,..because Branton and others are in their biggest part..from the other one people…and then we are lost…I HAVE MY OWN WAY,INSEKT (AH! AND KURS ALSO…INSEKTS OF COURSE…HEHE!!!). SO go for it!!!,lets write NEXUS that will be another volume XENUS (so, if oldest discoverings and the age of some aliens is very much older we just can imagine…and asking ourselves if we are products of a genetical operation or if we are originals (im sure from this second affirmation,we are 100% originals,and we came here and LATER lets happen all genetical operations,as Vlash with alamut Kaar at Russia, and they are practicing sine many thousand years ago, close to Greys with their time machines turn back time and are changing human being AND from very past times, (and important historical characters).Then, what is Greys techs from underground bases and genetical experiences from Vlash Gvoernments and Karistus?..In what point they take cccontact..AT ME!!!...At my brain is the key ,at Epsylon Eridani, in a ecological catastrophe that happened, that just happened or will happen, so this is Earth Planet,(or we are Eridanians in the future and we are sending messeages to ourselves now at present time…ver barnton language…BUT THAT'S ME!).,with our trees, with star chart travels,,and this catastrophe possibly was by reptilians (attacking Epsylon ecosystem) and erasind with their trees.And they want the history repeats...here!!!.And they giving an advice by that.Close with intraterrestrials.This "open" Planet (howl) as all planets do (so our sun,protects life,and whole universe life).I can get in touch with Vlash with reptilians,and their symbols to detect them,and genetical stuuf lineage,is very interesting, form the hybrids,light eyes and all this…everything touches with me!!!. I must consider very much to Dantes information or at least a part form my own has experienced all this, (with Laura,my Mother´s children). While Eve is Undergorund Bases main part,clones,and montaul technological greys,and 3-alternative.Maybe and that's only perhaps exists something close to both explanatory theories from a completely different point of view,that I ddint discover yet…or maybe I did!!!. And the Maynas also,which maths I learned with Argüelles, must be inside this main and central explanation or a more general vision point of view ..and with Epsylon support,intraterrestrials,and Maya.[know Epsylon,Intraterrestrilas and Maya and I knw Vlash,Kurs,SDA,…but I miss the point!!!!..something is wrong!!!...hehe!!!..And only Epsylon or from Epsylon we can find the meaning of everything of our species…
AKART---? Varginha?
(Photo)

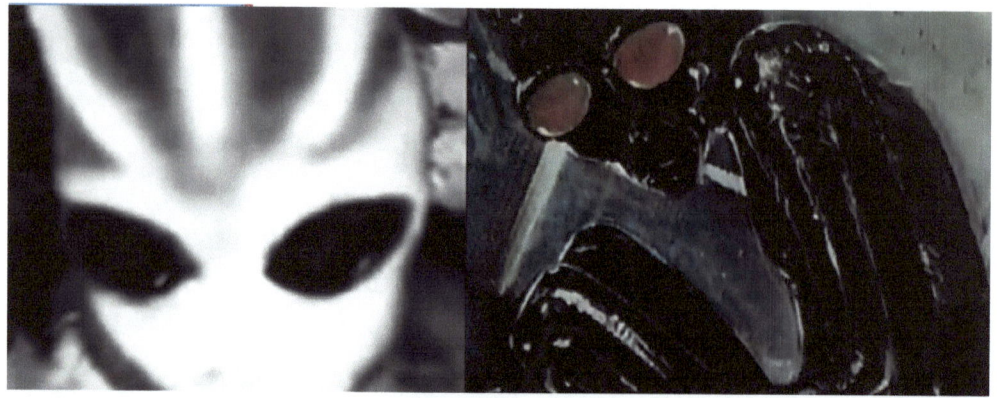

And Patricia…hehe!!!>
Vlash people are with gipsy people mess with them,and with Dhor Krystil ,and with wolvie hybrids…hehe!!!>
And the Torres Ufo we saw were form Akarts…hehe!! [That's mine].
My own documents:
-Porto Alegre Ufo
-Argentina Ufo
-Torres Ufo.
+ take youtube links
To crash inside this book.
What the stoyry behind the Akarts?.

Video photos: Akarts Blak eyes kids-Varginha.
AKART VS GOVERNMENTS
BRASIL VS USA (BRASILIAN ARMY THEY KNOW THE TRUTH AND THEY ARE COLLABORATORS WITH OLD Vlash agendas and new Grey technologies!).
27) I go on with Alien clasiification (a entomology or xenoentomology) new alien species…the ORANGE EBEN..[]
TO GET IN TOCUH WITH A.J.Gevaerd.
And what is my place in all that and with the rest of the world?.
Theres much more here involved you imagine[epsylkon says]
And the weather? How it gets modified with thorugh our feelings and Akart Technologies and Vlash SDA technologies?>
Both Sides…apparently they can be good or evil,but all is messed in such a way…that good can be evil and evil can help(well, is more difficult) which is our chriterion and that's my moment..me and Epsylon and Akart.
Akart protecting natural ecosystems

And Vlash improving artificial ways of life.
28) Can u "create" new brain cortexes?...hehe!!!!

.Stenonychosaurus'Dinosauroid'

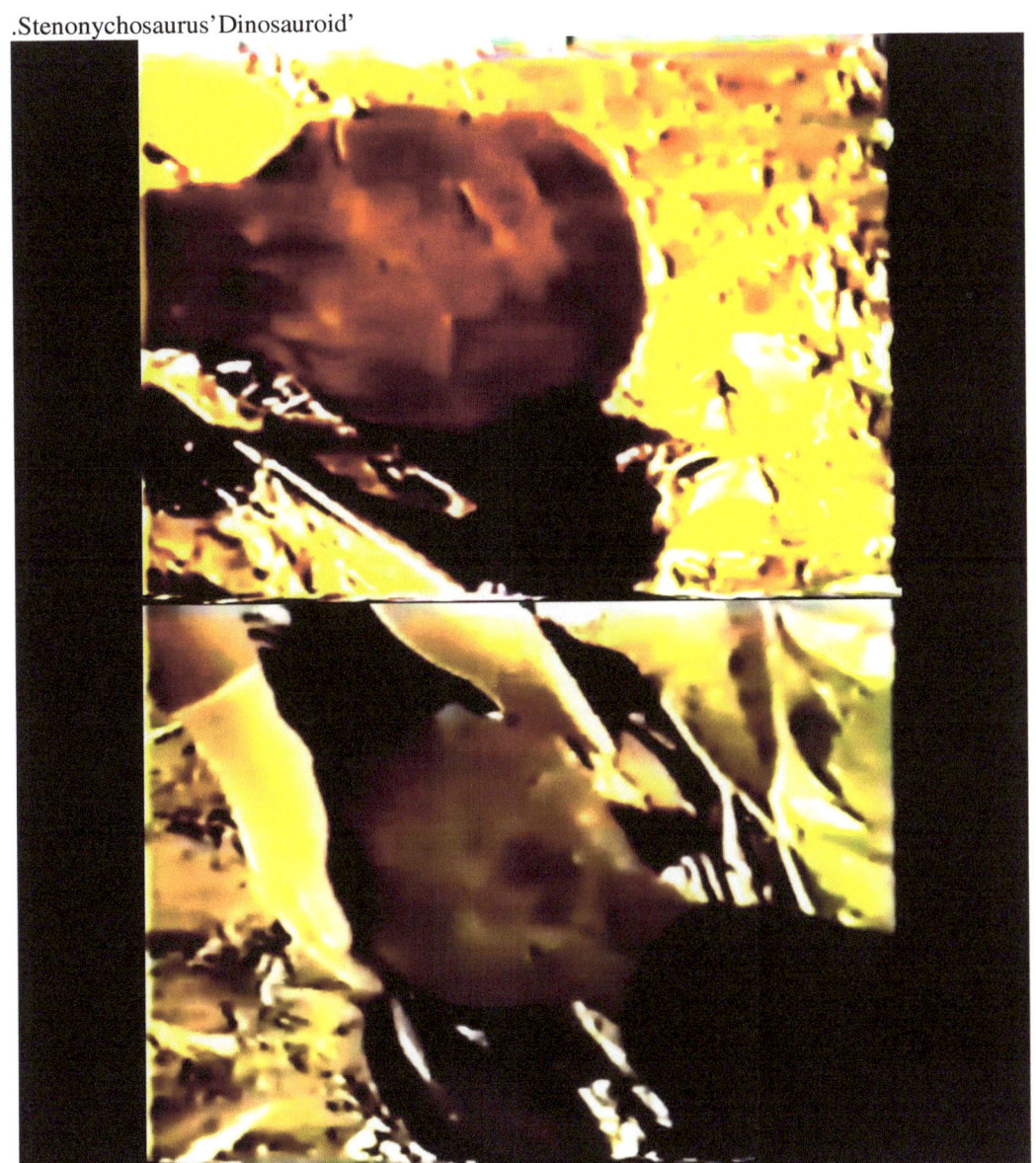

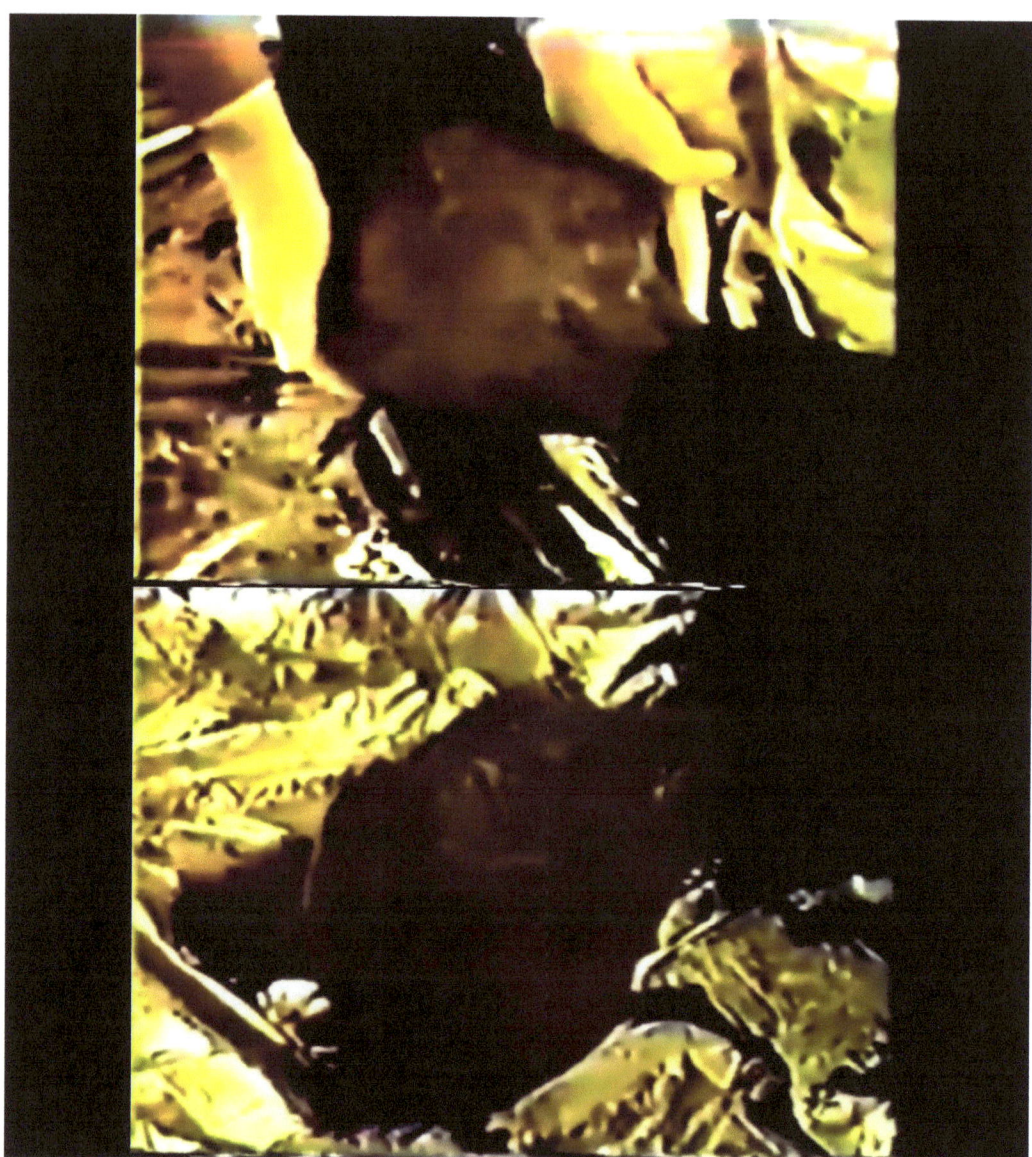

#NOTAS:

-We made it! .We have decoded the Machine, the Hive Alien-Extraterrestrial Mind,The Reptilia-Grey Empire!.

-It´s not only a mechanical entity,but an alien-extraterrestrial concept.

-Those who are growing spiritually have been chased bythe last remmants of NOW-Alien Empire.

-"And Love
Will make down
And up at new essence,
Together.
It will follow us
By eternal game
To the corridors of death
And Fateand once
We get there

Similar to floral centers
They´ll shine
Our souls
Because it has been
 Never been
No other thing".
-The Moon is not spinning around its axis,that´s the reason we can only see one face,and is the only one satellite that can do that,when I was a child they told me that she was an artificial satellite,I never understood very wellthat concept,but I see already that´s totally artificial .
-The Domes I Drawed at 5th EGB (Elementary School at Spain at 80´s) were real!,and there were 5 billion aliens living inside there,with gardens,swimming pools,..everything the needed with 13miles high!,as my drawings!!!.
-That´s the reason they want to get us into prisons and FEMA camps because they have their places at underground bases,the ellites have their ticket,how? Selling their brothers and sisters all around the world..all the governments and countries of the planet..The Resistance Goes On!!...hehe!!!.
-The Resistance has two main goals: To destroy mentally and spiritually hive alien mind and to free to human race all tech elite has and with wich all is possible…A tech given by the same aliens to governments to slave us, we, The Resistance uses to destroy them..as Virtual

Reality…Thank You!!!...hehe!!!.

The little Duck´s Bath..The Symbol of The Resistance…hehe!!!.

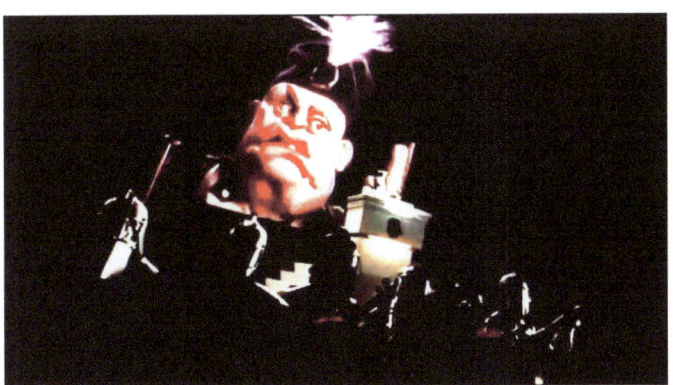

-Spirit has nothing to do with religions.It has to do with transparent technologies more than with invisible black magick tied-Technologies.

-Yes!,but these names are not Hebraic,they seem Aegiptyan..or Acadian,from Sumer..

-Anunnaki?.

-Things can be very complicated from now on…

-I understand that the Anunnaki Gods were both reptilians and humans..

-What you say?.

-Inside Reptilian Empire there could be some human "infiltrations"..Similar to known Elohim that help their brothers and sisters to get free from Annunaki-Reptilian programation,and be Human Beings,without no reptilian creation, we were humans since the beginning!!!.The repts were evolved from dinosaurs to twol ledged reptilian-humanoid and reptoids give them the shapeshifter possibilities..something was wrong with reptilians and had to get similar to human ,this is a human planet…

-I think you are right, so humans won the war for planet earth but with a little problem,the Annuanaki-Reptilian infiltration..

-Yes, this planet turns a kind of testing place for humans,because of this rept infiltration,a kind of frontline between the two opposite forces of the Universe…til now

-And religions want to program human beings to accept reptoid slavery of humans of all kind…To accept this situation of testing,but that wasn´t the truth..

-God never wanted to slave Human Race.And Sending the messengers tries to wake up humanity from this reptoid programation,but Reptilians transform them into religions to Mental and Spiritual slavery to humans,and they were chasing their own agendas to completely conquer this world,til now…

-They get allies of governments and create Reptilians-blood-lines to make possible this control from an Ellite of their same blood and nature.

-Do you believe shapesifting technology was in our planet since the begiinig?.

-Of course!.If you read the bible youll see many details of this tech and anothers.But God has His own Plan and send His Messengers to Earth from time to time..

-Nor religious leaders!..

-No, but human astronauts with some kind of technology to help human people to fight against repts.

-Advanced humans form another planetary systems..

-No, the same humans as you and me, but with advanced technology came here and try to explain the situation before reptilians catch and kill them…Nor a religious message but a technological message to attack invisible black magick reptilian technologies,…As transparent technologies inside their words…

-Aha!...I see…So into religions there were rememberings form these brothers and sisters from the stars…But with time Religions were reptilianatazed"..And now the only way to understand original message is breaking religious rules and get the intentional primitive technological message.

-That´s the truth…hehe!!!.

-Wow!!!...I see…Yeah!!!.So, I already see that some films,and video games are more usefull than clasiclal lectures…To get to The Source or God.

-Well, some new books are in that direction,you know..But yes, that´s it,to watch some films is more "teological" than read the Bible,brother!.And play video-games ,of course, that´s it!!!.

-But New World Order tries to rebuild religions once again to keep millenarian control over human population..

-Yes, included New Age,don´t forget and other similar "spiritual" branches..We need to evolve our spirituality but from this "transparent revolutionary technology" against the traditional spiritual black magick reptilian technologies..

-And the Grey technology..

-Its another part of the strategy from them,to get by their control all human minds in this world.You know all this gadgets tht everybody has to buy and know to control,..and the next Virtual Reality technology ,but humans can revert the situation,The Resistance can get these technologies from the Human Ellite in touch with the Reptilian-Grey Empire and give them to the benefit of all human kind and heal our world.That´s another part of our mission…hehe!!.Evrything we need we have already in this planet,do you understand?.

-Yes, Sir!!!...hehe!!!.

THE REBELLION AGAINST THE "GODS":

-Tecnocracy and technomonarquy are feudal type system...

-hey are fascinating arent they?(photos of nazi base) Is this a base from 1942 , established using Haunebu 1s and 2s (large nazi flying saucers/ moon shuttles) from their secret Antarctic base?there even appears to be a space capsule on the landing /takeoff platform in the middle of the swastika, indicating a base that was still in use.I believe that to have a base on the moon you must first have permission from the e.ts who live within the moon.This is why the NASA Apollo moon programme was terminated , because the e.ts in the moon , could no longer tolerate the 'space junk' being left behind and also they probably didnt want earths populations to know of their exsistance also.I theorise that secret treaties must exsist between the human fascists and the N.W.O elitists of earths secret space programme (the military space programme).These treaties with the moons alien inhabitants , allowing for the establishment of rh negative , human secret aristocracy controlled moon bases.

-"Artificial" races that you must know,so,i thought "natural" enviroment was parto f our planet ,and the rest of the Universe also, but I was wrong,many races and planets are totally antinatural,and that´s "natural " for them,.They are totally involved with a no-natural idea, and appart from "The Source "(as Vlash race) and its Soul-Matrix,and had broken all links or relationships with its natural soul-matrix (if they had),they are not more part of the essence of the Universe and its only purpose is to fight against those who try to defend them from conquer and destruction,the only way for those who are condemned ,without no hope,…,we called them fallen angels,demons,now we have enough tools to define them as artificial alien races,so AI at

our planet is by itself part of an Artificial Alien Race,as a way of conquer,from Reptilian-Grey-Vlash Hive-Mind taking another world and slave us as the rest of conquered worlds.
-I don't understand.So AI and Technosfere and all technologies are part of this Alien Invassion?.
-Well,not all technologies.Most of them,those related to virtual reality..yes.
-What kind of Alien Race?.
-Good Question,one of them are a synthetic Alien Race, Al Bielek prophetisated for 2011,and his discovered "wingmakers",those who defeated that race in other parts of the Universe.
-I got it!.
-But that´s not all.As you must know,"artificial" or "natural" are not exactly terms.We must be precise and saying "those who are connected with the Universal Soul-matrix" and those who don't,we can call them "T.W.C.U.S.M." or "CONNECTED" and "DIS-CONNECTED" and are always looking for energy-fuel-supply-feed goig on with its existence ,that´s all I have to say.
-Sorry,but you wanted to tell us something else,what´s it?.
-…,yes,there´s something else.You can find these energies at ALL environments,that's I want to say,so, you can find artificial presences at the most natural environments.For instane, I live at Brazil and I can check it here very precisely.When I detect artificial presence at this place is something obvious,then I realize who or what they are,beause they don't have any need to get hide here, they are very extreme at their manifestations,as well…And I can have a very good collection of all of those Artificial Alien Races,their weapons and new ways to conquer.
-Thank you .
-You´re wellcome.
Adds:
-1)their strategy consists at get all human kind under isolation,inside our houses,or mental isolation,or spiritual isolation to get us away from our essence.That´s their main strategy…hehe!!.And the governments help them in that goal.Laws,police,judges,try to inoculate us prisons,…mental prisons…virtual prisons..or physical prisons…Do you know what I mean?.
-Yes!.
-And what about Vlash?
-Well I knew them very well,they are totally un-natural,I had to do I very though effort to take away their vibrations,..
-And?
-Only with a conscient vow to the Source,is the only way..everyday…hehe!!!.
-I see,constantly!!..
-Yes!.
-(Un candidate a la presidencia de España que no es lo que parece-34 baños de sangre para el desde venezuela y los angeles de Epsylon ..Nexus Roma..y mi familia y la de Patricia…hh.
-24ª.AC. llegaron naves ..
-Filosoraptor
-Jesus was built and same purposes of someone else.
-Las nubes no son nubes son naves camufladas (negative).

-interacciones amistosas y no amistosas de la interaccion alien-humanos a lo largo de la historia 24 A.C.

-Bajar "BARAKA".
-Salvador fernandes Zarco-ristobal Colon..SDA hybrid?.
-Lushifar-Halverthy
-Mithilae trying to kill lushifar at erath during xv-xvi centuries
-Brasil tiene la function de colchon (200 millones de personas) del sistema un lugar de retiro para reptilianos..un lugar que pronto va a ser cambiado radicalmente...hehe!!!.
-Explosiones en planetas ceranos a la Tierra durante los siglos XV-XVI-XVII eran batallas entre enemigos galacticos.

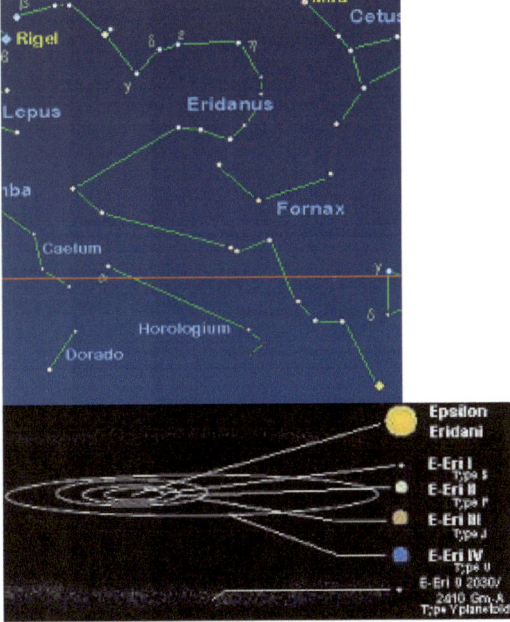

"EPSYLON X"

"Tras una necesaria pausa, regresamos!!!":

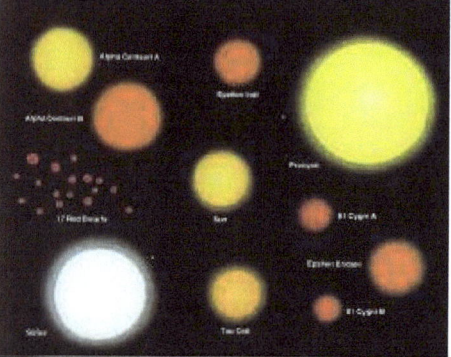

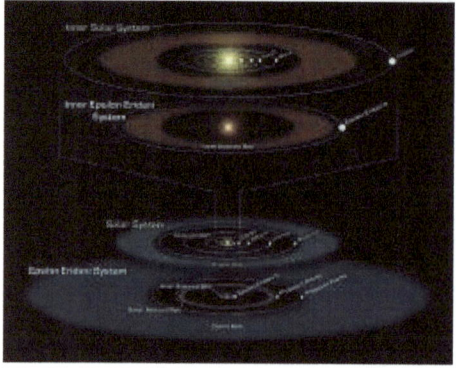

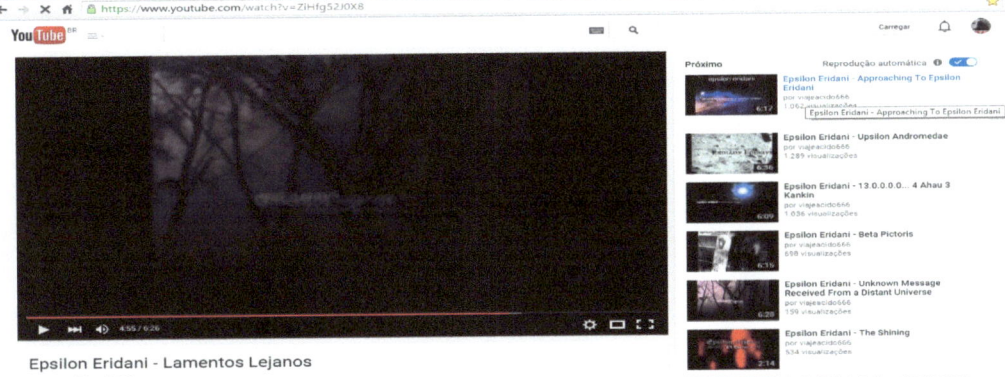

Epsilon Eridani - Lamentos Lejanos

viajeacido666

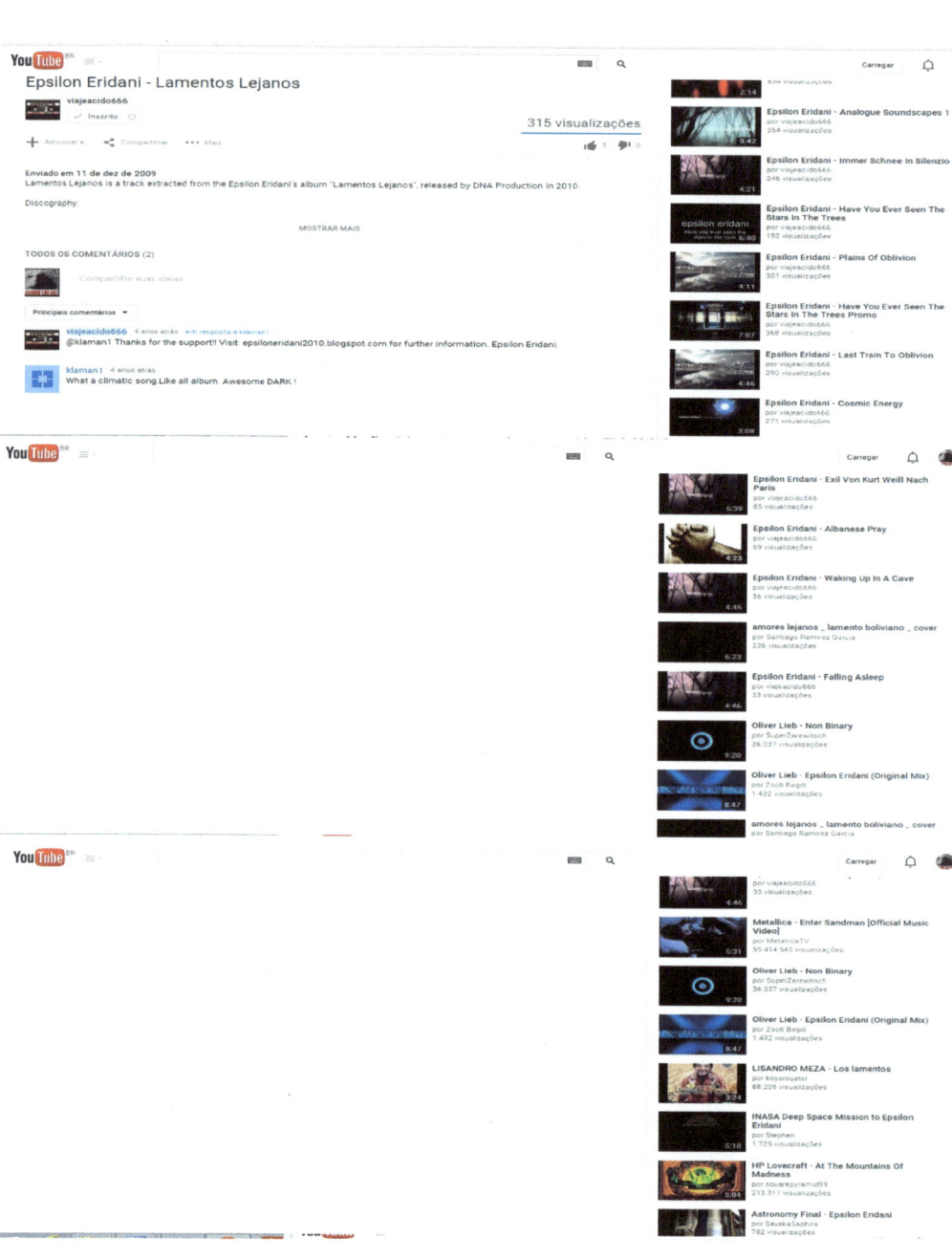

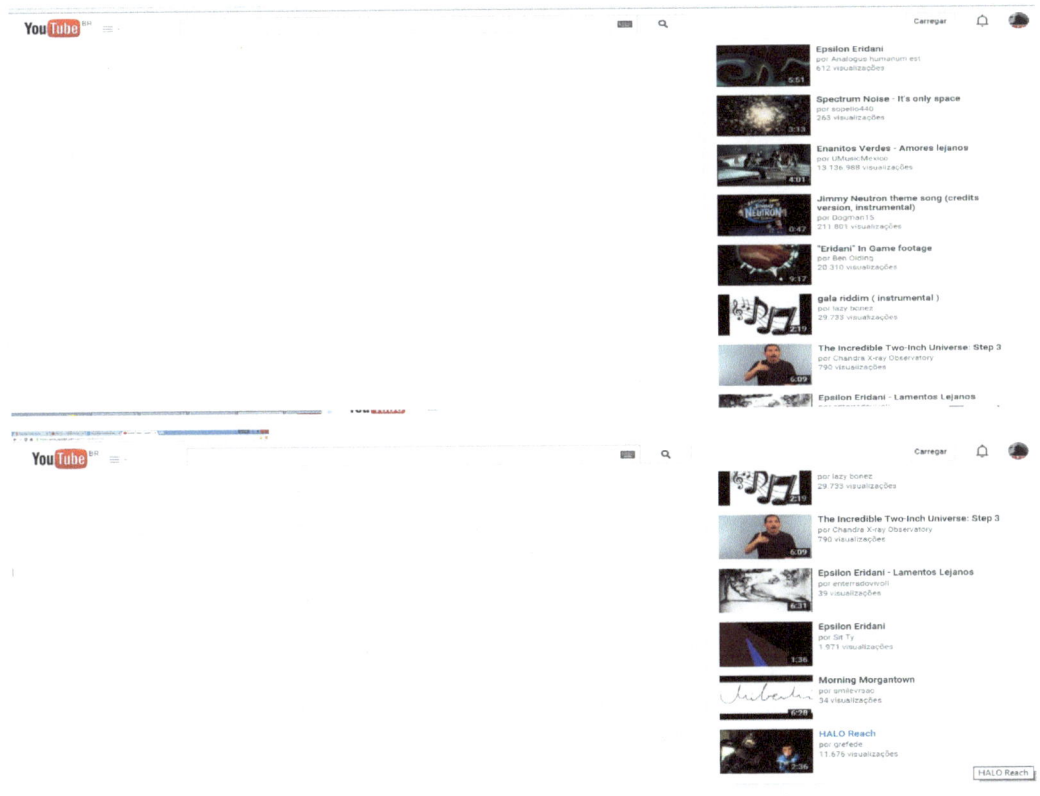

"La sincronización de la noósfera planetaria con el programa evolucionario estelar marca el advenimiento de la próxima era geológica, la Era Psicozoica. La Era Psicozoica se define como la secuencia normativa de la evolución de la super consciencia hiper-orgánica de una noósfera unificada telepáticamente." Las Dinamicas del Tiempo,6.1 que enlaza con el libro "La Tikrazia Insekto-La Geologia Neural y el Siglo XXIX" donde una serie de seres humanos transfieren su consciencia a la geologia karstica, montañas, sedimentos antes de la desaparicion de la raza humana y cómo tras su desparicion miles de millones de años después cómo las seres humanos surgieron desde las montañas en forma de una sustancia llamada"ganga" hasta tomar de nuevo formas vegetales basadas en el carbono.

#Señor! Los soldados delfin se aproximan juntoa las columnas ballena!!!

-Ya era hora, gracias al Único…hh.

E...ESE FUE EL COMIENZO...LA LETRA E...PORQUE?..de nexus-xenus
Estamos en (ir pegando todos los guiones de zona 84 aquí y construir el libro en base a
ellos…hehe!!!)…como estructura…huahuahua!!!

http://www.mp4upload.com/m7lwjt0343q1

[URL=http://www.mp4upload.com/m7lwjt0343q1]20150901_175425.mp4 - 267.6 MB[/URL]

No hace falta que sea una construccion lineal…hehe!!!.
Hay personas que les escuchan (casi como una técnica) a los reptilianos (annunaki),no como
voes en la cabezxa sino como seres reales.Yo escucho a los insekto y a los humanosd
originales,que estan relacionados con los insekto (kurs),como una segunda creaión (la primera
creacion fueron los hombres de madera,de paja,de metal de las leyendas Mayas o sea los
insektoides,etc…que no funcionaron y luego surgieron los humanos,como 2ª creacion frente a
los renegados de la fuente,(NATURAL).Yo esucho en el plasma a los Arkontes (amaebas,y
anunnaki tambien) y sé lo que nor ordenan o nos quieren ordenar y tambien escucho o soy apaz
de escuchar tambien a los humanos ancestrales y alos insekto(kurs) y ellos (los arkontes,anunaki
y demas   NOS CULTIVAN).
NOS CULTIVAN!!!.Y con el tiempo he logrado diferenciar primero acentos,luego
idiomas,luego lenguajes y he podido distinguir sus agendas,sus planes,sus puestos de
observaion,de combate,sus unidades de ratreo,sus leyendas,sus linajes.Es por eso que he
"escuchado" y he leido en mi mente a los Anunnaki,a los Arkontes,a los infiltrados no omo
energias inorgánicas sino como seres reales que nos proyectan sus órdenes en nuestras mentes
(o lo intentan) y luego a las otras razas positivas (pleyadianos,intraterrestres,..arturianos..) que
intentan limpiar ese canal que es nuestra mente (de la programacion o del CULTIVO
MENTAL..ya que los anunaki "siembran" sus formas de pensamiento como virus o como
baterias para que se reproduzan en nuestro organismo y se apoderen completamente de el-
mentalmente) y dejarlo vacio.
VACIO es la naturaleza de la mente,para y que conecte con la fuente,entonces surgen todas
nuestras comunicaciones (si queremos) con nuestros hermabnos y hermanas pero la
comunicaion fundamental con la fuente es el silencio básico,el sonido del vacio elemental,el
ruido la confusion de lenguas,la cacofonia es producto de estas otras razas "infiltradas" pero que
son mas un acidente que una realidad dominante.
Nosotros somos originalmente humanos y todo lo demas son fuerzas de coercion y coacion (que
aparentemente o en prime rlugar son atrayentes,pero luego dejan pasar el charme o el glamour a
la mayor de las

Opresiones).Llegar a la liberación es un trabajo de limpie-za zen o zazen..hh..limpiezazen.
Este extracto es la parte fundamental del libro.

Como fui descubriendo leyendo en mi propia mente toda la historia cósmica o exocosmobiologia,ésa es la realidad,recuperar nuestro legado autentico,tras quitarnos el "aparejo" artificial, (O MATRIX)—luna sic.

Que es historicamente genetico pero ahora es realizado tecnologicamente.mi fuente de informacion es la propia fuente que me envia toda la informacion(sic) a mi canal vacio,o abierto,con todo detalle,por eso todos somos contactados,porque podemos recibir de forma natural todo lo que somos,lo que fuimos,lo que ocurrio y lo que seremos.Las ayudas y el gran plan divino,pero que al ser natural forma parte de nosotros y nos parecen que no existe.Pero al tomar consciencia de estas injerencias el gran plan divino entra en funcionamiento y se hae CONSCIENTE (antes era inconsciente),y es por medio de la exocosmobiologia que se nos hace aparente a mi, incluso esta smismas palabras.Que llevan dentro de si todas las tecnologias,las armas de guerra,los bombarderos,y el color de nuestros ojos,las respuestas en nuestro color de piel,erc…y luego las respuestas son confirmadas exteriormente o conscientemente (a traves de canales de la "realidad") pero la conexión ha sido realizada ya in-conscientemente o sin vacio (sin silencio) y ya conocemos naturalmente TODO,luego unos pocos o por circunstancias excepcionales todo esto se hace consciente .Lo que está pasando es que al abrir los anales de unos pocos se han abierto los canbales de TODOS,y entonces la Cacofonia o los ataques de los aliens regresivos aumentan!!!.Cual va a ser el resultado:? Nadie lo sabe,una civilizacion absolutamente cósmica y exhuberante como nunca se ha visto antes …HEHEHEHE!!!.
Guion de la casta de los metabarones…para "Nexus"…hehe!!!.
yagusari---palabra de gran poder potencia para vencer a todos tus enemigos…hehe!!!>

#HAY MUCHAS RAZAS GRISES…UNAS SON AMISTOSAS Y OTRAS SON REPTILIANAS…HH

BAJAR METABARONES Y VER LOS COMICS EN EL JATDIN PARA REDACTAR "NEXUS" MUY EN EL CONTEXTO DE LA VALENCIA DE 1984 CONTEXTUALUZ EL LIBRO AHI Y EN LA VALENCIA D PEREZ CASADO PERO SIN PERDERTE EN LOS DETALLES HISTORICOS SINO EN LA PSICOLOGIA DE LOS PERDONSJES MAS COMO UNA EXCUSA QUE COMO UNA INTROD. HISTORICA Y HAWAIKA...COMICS METAL HURLANT...Y ESCRITO EN BRASIL...COMO AHORA EXORZIZAR LOS MOV REPTILISNOS DE AHORS MISMAS ARMAS METAPSICOLOGICAS...AH Y LO MAS IMPORRANTE RECORDAR QUE ESTAMOS EN UN AÑO ABEJA-ARTE (GAY) Y QUE ES UN AÑO NO PARA CAMBIAR NADA SINO PARA CREAR..Y LEER OTRA VEZ "EL

😊 😊 😊 😊

GRAN PLAN DIVINO" OBLIGAO!!!... .SIN LIMITACIONES...HH..PLAZA DE OTROS ESTACION DEL NORTE MALVAROSA DISEÑADOR DE VALENCIA MONTESINOS DISCOTECA DE LA MALVAROSA...NOMBRE...HH PERO VOLVER AQUI PARA VOLVER OTRA VEZ ALLI Y CAMBIARLO TODO LA HISTORIA...HH Y DE NUEVO VUELTA AHORA A LA GALAXIA LYRA CON XENUS....Y BRASIL 2015 EN RIO GRANDE DO SUL ESA TEXTURA...HH...METER PROTEUS EN VALENCIA...HH...EL AÑO 1981 FUE OTRO AÑO DE METAL...HH EL AÑO K VIENE SI SERA UN AÑO DE VIAJES MUCHOS VIAJES...HH
jEXPERIENCIAS GENETICAS REPTILIANAS...NEXUS...FOTO...#geneticalreptiliansxxxc en LENOVO.Hum
PONER UN VÍDEO DE YOUTUBE DE LA VALENCIA DE RICARDO PEREZ CASADO
#HABLAR EN "NEXUS" DE LO DEL PRIMER MINISTRO GRIEGO.
Y SI CLON...INCLUIRÑO Y LO DE TRIX Y LOS EMBARAZOS DE 3 MESES....HH
Alain Villeneuve...hh
#HABLAR EN "NEXUS" DE LO DEL PRIMER MINISTRO GRIEGO.
Y SI CLON...INCLUIRÑO Y LO DE TRIX Y LOS EMBARAZOS DE 3 MESES....HH
#EN LO QUE OCURRIO EN EL DESIERTO DE OUARZAZAT LAS RELIGIONES NO TIENEN NADA QUE VER...NINGUNA...ESTO QUE ES LA REALIDAD ESTÁ MAS ALLA..DE LAS RELIGIONES DE TODAS...CLARO!!!...HEHE!!!.
Y ES LA REALIDAD...ES ALGO QUE SON EMBARGO ESTA TOTALMENTE ENRAIZADO EN NUESTRA ALMA EN LO MAS PROFUNDO Y AHORA ES EL TIEMPO DE SACAR A LA LUZ ESA HISTORIA ESA LUCHA...Y TU ERES PROTAGONISTA SOPHIA TU ERES LA HIJA DEL DESIERTO...NO LO OLVIDES...HH
#SUNSHINE HAVE BLOWN MY MIND AND THE WIND BLOWING MY BRAIN...HH
#M.I.A.B.23:00MECHH.TMR.G.O.#
THE SALVATION OF THE UNIVERSE...THE RETURN OF AL ANDALUS (A.A.)...HH
#ESTOY EN EL PAIS DE LOS ZOMBIES VAMPIROS...HUAHUAHUA!!!

e - Commander X).
"Our explorer J.D. (name on file - Commander X), who is a mountain guide of the Mystery Mountain near Joinville (where there is supposed to be an entrance) said that, several times, he saw a luminous flying saucer ascend from the tunnel opening that leads to a subterranean city inside the mountain, in which he heard the beautiful choral singing of men and women, and also heard the 'canto galo' (rooster crowing), a universal symbol indicating the existence of subterranean cities in Brazil. He said that the saucer was so luminous that it lit up the night sky and converted it into daylight. On one occasion he met a group of subterranean men outside the tunnel. They were short, stocky, with reddish beards and long hair, and very muscular. When he tried to approach them, they vanished. Often he saw strange illuminations in this area at night which were probably produced by flying saucers (We use the name 'Mystery Mountain,' rather than reveal the true name of the mountain, so that unwanted outsiders will not come here to locate it). Throughout my many years of research I have accumulated a vast amount of data which would indicate that these entrances to subterranean cities abound throughout the region.

"An elderly man living in Joinville once told me that he had visited a tunnel near Concepiao in the state of Sao Paulo, and saw in the distance a marvelous subterranean city with vehicles darting back and forth, evidently traveling through tunnels from one subterranean city to another.

"Although the following report requires confirmation, it was told to me by an explorer named N.C. who said that he had visited a tunnel near Rio Casdor and had met a beautiful young woman appearing to be about 20 years of age. She spoke to him in Portuguese and SAID that she was 2,500 years old. He also met a bearded subterranean man

(Note: Often humans encountered in aerial disks or subterranean caverns declare that they are extremely old by humans standards. On the surface this might sound next to impossible, unless a revolutionary scientific breakthrough on the part of these human 'aliens' has allowed them to retard the ageing process to an extreme degree, or could the possibly that they are separated from the degenerating radiations of solar rays explain their allegedly greater longevity? Another possibility would be that throughbionics/biological transplants/prosthetics, etc. the lifespan of human beings possessing advanced biological and technological sciences might theoretically be increased dramatically. Incidentally, the writer and traveller Robert Stacy-Judd in some of his booksdescribed an exploration he and others in his party made of the peripheral areas of the Loltun caves of Yucatan.

Legend says that at least one group of people, fleeing persecution, entered en masse into the massive Loltun caves and were never seen again. Stacy-Judd tells of his own encounter with a 'cave hermit' deep in the cavern chambers who claimed to be well over 1000 years old, and who said he was a guardian of the cave and of the treasures--and city?--which lay deep below in the unknown depths, 'unknown' that is, except to the strange 'hermit'. Aside from photographs of this hermit which appeared in some of his works, the author also revealed photographs of 'underground gardens' consisting of areas of the cave which contain small patches of 'jungle', watered and lit through parts of the cavern ceilings which had collapsed, exposing them to the outer world. Whether such claims of longevity are real or whether the "subterranean" people were just playing with the minds of such explorers who encountered them, is uncertain - Branton).

2.El origen del hombre ya fue de por si bastante "biomecanico"..Fue producto de una idea de bioingenieria tan arriesgadaque cambio totalmente el propio Universo,somos un capricho…Somos en realidad un producto en si de bioingenieria-biomecanica-biorobotica del Cosmos, de La Fuente,y como tal fuimos originados como una especie de sres robóticos orgánicos,cuya función todavía desconocemos, estamos buscando…Los regresivos intentan imitar esa creación original pero su hazaña es futil, no pueden igualar la perfección de La Fuente [The Source].Éso es lo que estamos viendo en nuestro planeta en los últimos tiempos,el desvelamiento de los planes de los regresivos en crear una raza mutante humana-reptiliana,y desperdigar la propaganda de que somos descendientes de una creación de bioingenieria reptiliana o Anunnaki.Qué?..Para nada.Nacimos humanos originalmente,y somos perfectamente autonomos y diferenciados del resto de razas ,de otras razas,sobre todo reptilianas…hehe!!!.Sólo hay que mirar en Las Crónicas de la Guerras de Lyra y saber lo que ya ocurrió alli…corrupciones genéticas,disfunciones biomecánicas por infiltraciones de quien? De nuestros mayores enemigos…Los Grises,los reptilianos,los Vlash,Los Maitre..hehe!!!.La última trampa es a la que estamos asistiendo…hehe!!!.[Last Deception].

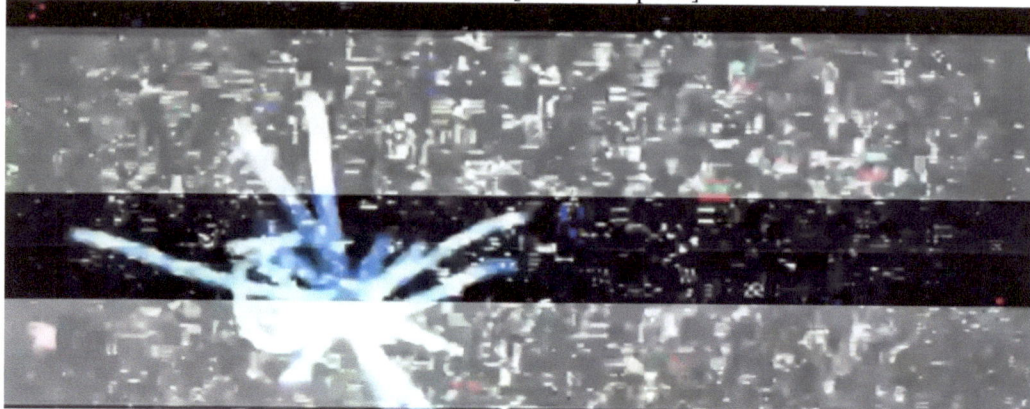

-Las nubes no son nubes son naves camufladas (negative).

-interacciones amistosas y no amistosas de la interaccion alien-humanos a lo largo de la historia 24 A.C.

-Bajar "BARAKA".
-Salvador fernandes Zarco-ristobal Colon..SDA hybrid?.
-Lushifar-Halverthy
-Mithilae trying to kill lushifar at erath during xv-xvi centuries
-Brasil tiene la function de colchon (200 millones de personas) del sistema un lugar de retiro para reptilianos..un lugar que pronto va a ser cambiado radicalmente…hehe!!!.
-Explosiones en planetas cercanos a la Tierra durante los siglos XV-XVI-XVII eran batallas entre enemigos galacticos.

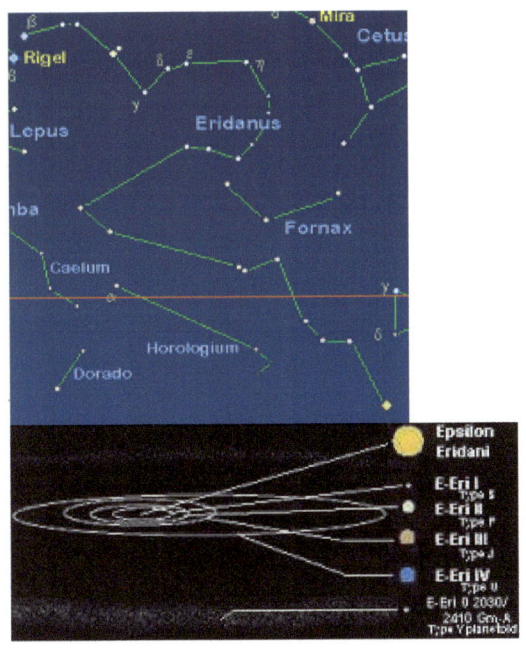

"EPSYLON X"

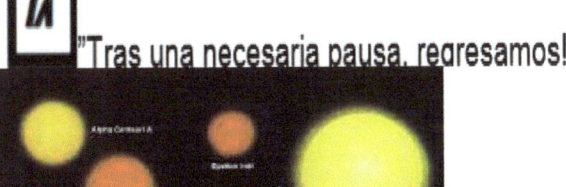

"Tras una necesaria pausa, regresamos!!!":

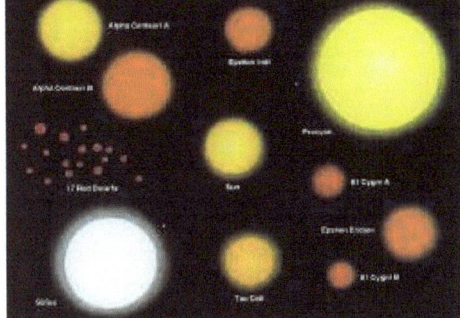

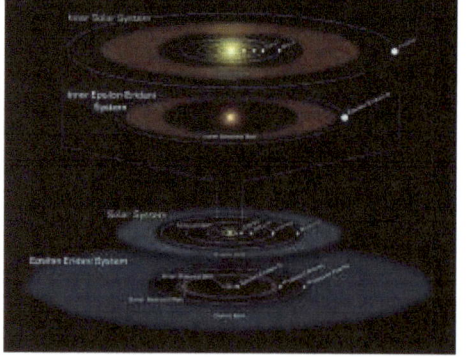

of top scientists and
engineers from all around
the world gathered together
and designed a deep space
probe sent to Epsilon
Eridani to transmit
high-resolution images back
to earth. These group o

1:20 / 5:09

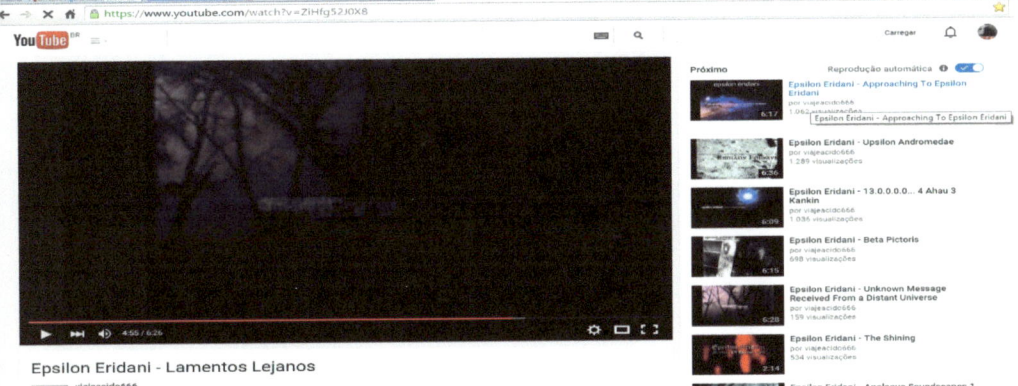

Epsilon Eridani - Lamentos Lejanos

viajeacido666

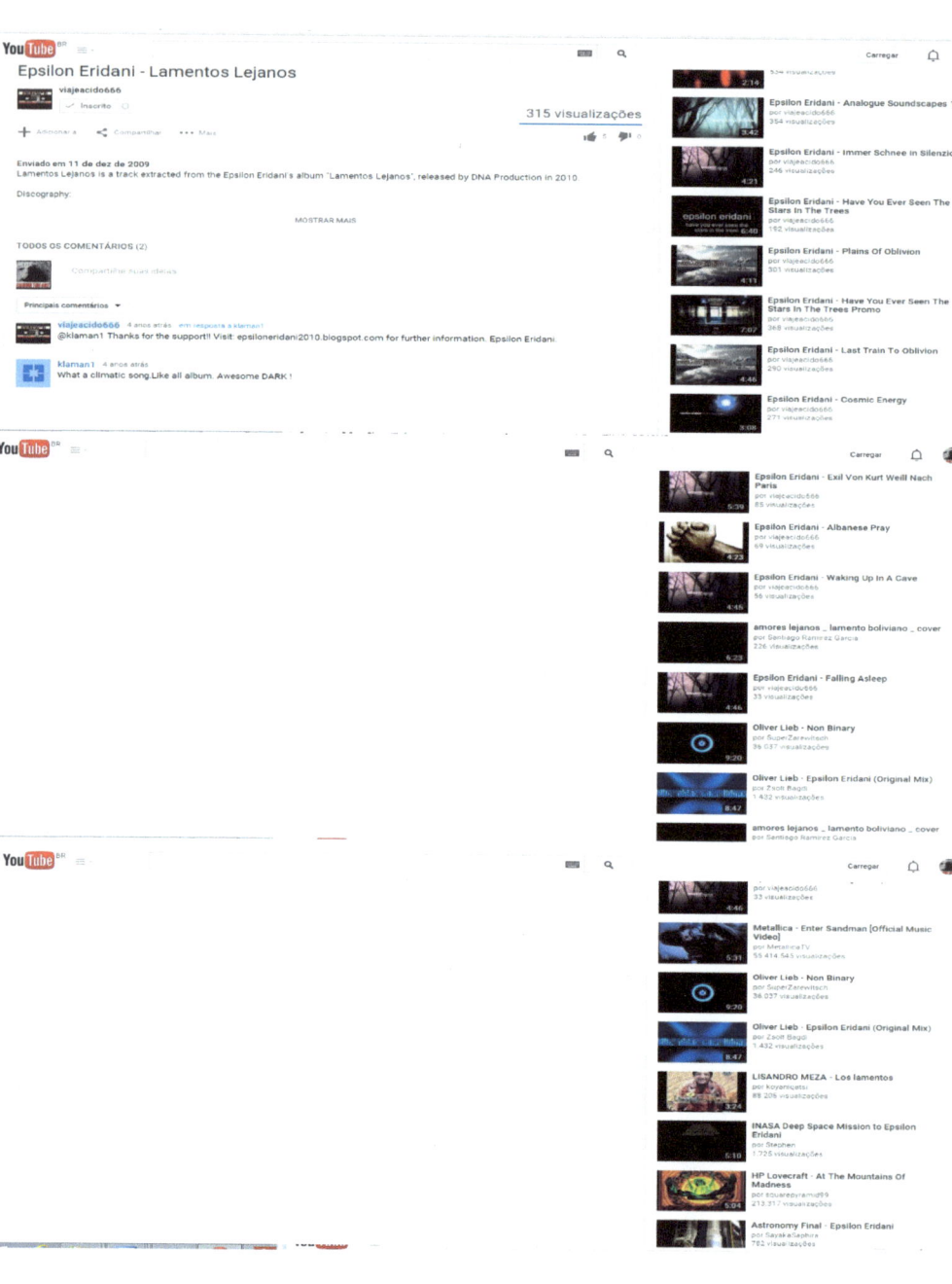

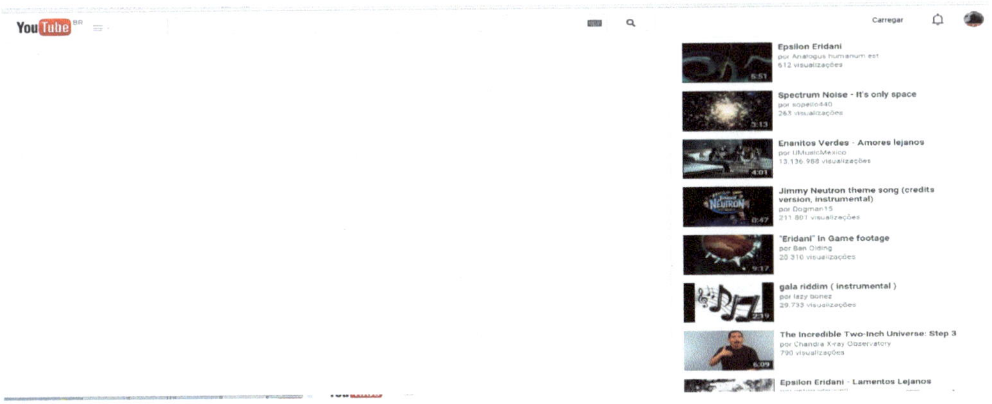

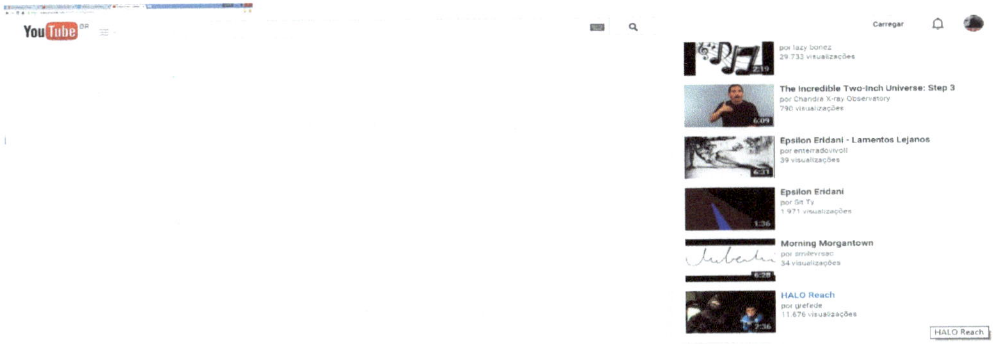

PREMIUM WORDPRESS THEME....HH

PREMIUM WORDPRESS THEME....HH

http://www.mp4upload.com/m7lwjt0343q1PREMIUM WORDPRESS THEME....HH

PREMIUM WORDPRESS THEME....HHAnyway, another incredible example of all this in the Cymatics videos is seeing almost:human-like figures forming from the particles when certain sounds are emitted. Our bodies are also the result of sound resonating

energy into form and if our minds are powerful enough to change the sound range of the body, it moves into another form or disappears from this dimension, altogether. This is what is called shape-shifting. It is not a miracle, it is science, the natural laws of creation. The full-blood reptilians of the lower fourth dimension can therefore make their 'human' physical: form disappear and ~ bring forward their reptilian level of existence. They shape-shift. To us in this dimension they appear human, but it's just a vibrational overcoat,

#DEFENSORES DE EPSYLON DEDICARLO A SOPHIA..RAPIDO

#SURPRISE ATTACK!!!...HUAHUAHUA!!!

#HABLAR EN "NEXUS" DE LO DEL PRIMER MINISTRO GRIEGO. Y SI CLON...INCLUIRÑO Y LO DE EVA Y LOS EMBARAZOS DE 3 MESES....HH

#grabar sonidos en el jardín...hh

#WHATS SDA?...HH SEGUN TETMINOLOGIA DE DANTE SANTORI..HH?"per áspera ad astra"

#TÚ ERES LA NOCHE ETERNA...Y SOLO SE RECUBRIO DE CARNE DE LABIOS DE UÑAS DE PEOR SI ABRIESEMOS PERO POR TUS OJOS Y TÚ CABELLO EL TIEMPO NO HA PODIDO CUBRIR Y SALE MOSTRANDOSE ..Y SI ABRIESEMOS PIR DENTRO DE TI NO HABRIS MUSCULOS HUESOS SANGRECSINO LAVPROPIA NOCHE OSCURA Y PROFUNDA LS ETERNIDAD DEL COSMOS HH...Escribir cuento..hh

#1:13:23m mec THE FLY INSEKTO-MECHANICAL MACHINE THE LAST FUSSION...HH

#ESTOY EN EL PAIS DE LOS ZOMBIES VAMPIROS...HUAHUAHUA!!!

civil war 5 ahora!!!...hh

#TROPAS GUARANI-KAIOWA SE AGRUPAN EN MATTO GROSSO DO SUL....HH..EL EJERCITO BRASILEÑO ESTA LLEGANDO PARA ATACARLES...POR EL PELIGRO QUE SUPONDRIA PARA EL ESTADO BRASILEÑO LA AUTONOMIA "DE FATO" DE GRANDRES AREAS DE SU TERRITORIO Y K PASEAN A MANOS DE SUS LEGITIMOS ADMINISTRADORES LOS PUEBLOS ORIGINARIOS TANTO EN BRASIL COMO EN TODA AMERICA Y EL RESTO DE ETNIAS DEL MUNDO...HH

AND THE NWO GOES ON DESTROYING PALMIRA BY ISIS HANDS WHILE CONDECORATES TRAIN "HEROES"TO STOP A STEAL ACT AT FRANCE...SAME SYSTEM, DIFFERENT CHARACTERS.SYSTEM HAS TAKEN PSICHOPATIC LEVEL AT FASTEST RATE..THE RESISTANCE GOES ON!!!..HEHE!!!.

#HOY HE SOÑADO CON EVA...IBAMOS A UNA ESCUELA EN RUZAFA Y TODO EL TIEMPO LA COGIA DE LA CINTURA...EL MEJOR SUEÑO DE MI VIDA..HUAHYAHUA!!! 2 09 2015 CAL GREG...HH

06 15 foto zen

#M.I.A.B.23:00MECHH.TMR.G.O.#

THE SALVATION OF THE UNIVERSE...THE RETURN OF AL ANDALUS (A.A.)...HH

#KUKITINHA!!!...HEHE!!!

#EN LO QUE OCURRIO EN EL DESIERTO DE OUARZAZAT LAS RELIGIONES NO TIENEN NADA QUE VER...NINGUNA...ESTO QUE ES LA REALIDAD ESTÁ MAS ALLA..DE LAS RELIGIONES DE TODAS...CLARO!!!...HEHE!!!.

Y ES LA REALIDAD...ES ALGO QUE SON EMBARGO ESTA TOTALMENTE ENRAIZADO EN NUESTRA ALMA EN LO MAS PROFUNDO Y AHORA ES EL TIEMPO DE SACAR A LA LUZ ESA HISTORIA ESA LUCHA...Y TU ERES PROTAGONISTA SOPHIA TU ERES LA HIJA DEL DESIERTO...NO LO OLVIDES...HH

#

#INFILTRATION AT PATRICIA'S GYM...HOW CAN I GET THIS ?...HUAHUAHUA!!!!...I SEE...I UNDERSTAND VERY WELL...NICE SHOT!!!

Alain Villeneuve...hh

#SUNSHINE HAVE BLOWN MY MIND AND THE WIND BLOWING MY BRAIN...HH

red social de cortos "vine"

#

HACER UN VIDEO DEL CACTUS REPTANTE DE PATRICIA Y COLOCARLO EN VIMEO O YOUTUBE O OTRA PLATAFORMA PERO K SE VEA O UNAS FOTOS...HH IMPRESIONARA EN EL FB...HH.

c412f538d825 contraseña wifi mío...hh
#EXOBIOLOGIA...EXISTE EL HOMBRE LOS ANIMALES LAS PLANTAS Y LA
EXOBIOLOGIA O SERES BIOLOGICOS ALIENS AQUI EN NUESTRO PLANETA..DE
HECHO LLEVAN AQUI MAS TIEMPO QUE NOSOTROS CON LO CUAL NOSOTROS
SERIAMOS LOS ALIENS Y/O EXTRATERRESTRES..O UNA RAZA MAS AHORA EN
PROCESO DE AUTODETERMINACION DEL RESTO DE RAZAS SIMBIOTICAS : :
#COJO LA BICI DE LA GUERRA CIVIL...CIERRO CON CANDADO EL CARCAMO Y
BUSCO A PATRICIA..A PARTIR DE AHORA LLAMARE A ESTE TELEFONO "LA
BICI"....HH.
#1:11:26 THE FOTO FLY INSEKTO THE KURS...HH

#HACER VIDEO EN MOVIMIENTO...HH

87 ALMANAQUE.
Hombre
Install Party 2015
Publicado el 19 Diciembre 2014
Como cada año, os invitamos a la jornada que organizamos desde el hacklab para iniciar el año.
Este año realizaremos un taller introductorio a Arduino durante la mañana, y durante la tarde
tendrá lugar la install party.
Los horarios serán los siguientes:
Sábado día 3 de enero
+ De 11: 00h a 14: 00h: Taller Arduino
Introducción al hardware libre e implementación de sencillos modelos electrónicos interactivos.
Habrá material disponible para experimentar.Recomendamos que traigas tu ordenador.
+ De 16: 00h a 22: 00h: Install Party
Instalación de sistemas GNU / Linux. Lleva tu máquina y Donal una nueva vida. Pasa un buen
rato jugando con una Raspberry como consola Arcade.
Lugar: Casal Popular la Turba (C / Nuevo nº12, 43800 Valls)
Escribir un comentario
Salvemos Internet
Publicado el 21 Febrero 2014
Nos piden la atención sobre la campaña internacional de apoyo a la neutralidad de la red:
http://savetheinternet.eu
http://savetheinternet.eu/es
La Neutralidad de Internet es un principio básico, uno de los fundamentos de guifi.net, y es
fundamental para las redes comunitarias descentralizadas, pero en Internet hay grandes
empresas que presionan pidiendo compensaciones económicas que pueden perjudicar la
capacidad de construir y mantener la red de redes que es Internet.
La campaña consiste en ponerse en contacto con los diputados al Parlamento Europeo, como
una demostración de que la neutralidad es importante en Internet para todos los ciudadanos
europeos y hacerles saber que es importante que voten a favor de la neutralidad de la red.
La semana del 27 de febrero hay una importante votación sobre la neutralidad.
Por favor, visite http://savetheinternet.euactúe y difundid la importancia de la Neutralidad!
Texto vía lista [guifi-users]
Escribir un comentario
Install Party 2014
Publicado el 24 Diciembre 2.013
Como cada inicio de año, desde Hacklabvalls os proponemos comenzar con una fiesta de la
instalación de GNU / Linux.
Para aquellos que queráis cambiar de sistema operativo en sus máquinas, o consultar dudas si ya
tiene un sistema GNU / Linux os esperamos el próximo sábado día 4 de enero a partir de las 4
de la tarde en el Casal Popular laTurba.
Este año realizaremos varias actividades en las que podrá participar:
- Jugar con juegos clásicos de 8 y 16 bits (emuladores)

- Sesión de corto-metrajes de animación Creative Commons(creados con Blender
#Y/O PUEDO ENVIAR X EL MESSENGER ...DEL NUEVO E-MAIL...HH
PONER UN VÍDEO DE YOUTUBE DE LA VALENCIA DE RICARDO PEREZ CASADO.
26 ZONA84
#EL MANUSCRITO DE JACQUES DE MOLAY DEDICARLO A ISMAEL...HH.
#LOS OJOS DE LOS HIBRIDOS DE LOS VLASH BRILLAN EN LA OSCURIDAD COMO
LOS DE LAURA LA MADRE DE MIS HIJOS EN LA DISCOTECA DEL SALER
CREYENDO QYE ERA SIRIANA A ERA VLASH O SEA SIRIANA B...HH..AÑO 1999
+....ESTE ES EL SIMBOLO...EXACTO!!!...HH
#antes del 3 de Agosto cal Greg tengo que recargar otra vez...hh
#1 32 tienda de discos...y 0:35 0:37 0 42 m mec 0 44 m mec
VIDEOS...HUAHUAHUA!!!:
3-conversaciones con un escritor
2-guarani-kaiowas Matteo Grosso do sul
1-los que van de folklore indígena...hh
I READ BEETWEEN THE LINES..SO MY BOOKS ARE VERY MEDITATED…THERE
ARE NO RANDOM OR MISTAKES…THEY ARE VERY SPECIFICS;;…WELL I LIKE
NOVELS BUT…HH.
#la auténtica vida es 'esta ,en el centro epsylon....hh
#HAY MUCHAS RAZAS GRISES...UNAS SON AMISTOSAS Y OTRAS SON
REPTILIANAS...HH.
#bajar el precio de los libros y comenzar el blog "el ejércuto de los grises..huahuahua!!!..y usar
el tor....hh.
#BUSCAR "THE 59 STREET BRIDGE SONG".
EXISTEN 20 O 30 SUBMETALENGUAJES O DIRECCION DE CONVERSACIONES BAJO
EL MANTO DE LA SUPUESTA REALIDAD CONSENDUADA ESTOS SON TODOS
ELLOS REPTILUANOS GRISES EN PAISES COMO BRASIL POR EJEMPLO LA
DIRECCION DE LA SUPRACONVRRSACION LA RSALIZAN OTRAS RAZAS POR
ENCIMA DE ESTE SUSTRATO DE CONSPIRSREPTILUANISMO Y SUPERA EL
UMBRAL EN LA CONVERSACION HABITUAL HUMANA AHI ES DONDE NOS
ENCONTRAMOS CON NUESTROS HERMANOS Y HERMANAS DE LAS
ESTRELLAS..HH.
#YA HE DEJADO DE TOMAR NOTAS DE "NEXUS"...HH HOY 27 /08/2015 CAL GREG.
Sesión de corto-metrajes de animación Creative Commons(creados con Blender.

¡EXPERIENCIAS GENETICAS REPTILIANAS...NEXUS...FOTO...#geneticalreptiliansxxxc
en LENOVO.Hum.
#BUSCAR SIGNIFICADOS DE FRANCISCA Y DE ARCO APELLIDO EN GOOGLE...HH
#LOS INSEKTOS Y JESUS ...O DHOR KRYSTIL...HH.
According to former 33rd degree Mason James Shaw, author of "THE DEADLY
DECEPTION", the U.S. headquarters of the Scottish Rite is located in the "House of the
Temple" in Washington D.C. and, according to some, it sits directly over an antediluvian
system of 'Atlantean' tunnels and ancient underground chambers called the 'NOD' complex,
which serves as a major NSA-Sirian-Grey center of collaboration. Some believe that the
antediluvian or Atlantean alchemists or sorcerer-scientists had begun experimenting with
elemental forces and that their experiments had gone out of control and created a temporal rift in
the space-time continuum in the so-called 'Bermuda Triangle' region, opening up a hole
between dimensions and leaving 'electromagnetic fallout' which has had adverse effects to this
day.
According to Des Griffin, Pike and Mazzini established the 22 Illuminati 'Palladium' lodges for
the express purpose of creating the right-wing Nazi and left-wing Communist movements and to
lay the foundation for three world wars which they hoped would wear-down the masses to the
point where they would accept a New World Order dictatorship as the only peaceful alternative.
From the alien perspective, the New World Order would offer easier control and massive
population reduction at the same time.

As I have suggested, we can also surmise that the 33+ Masons 'leading' the right wing factions and the 33+ Masons 'leading' the left wing factions are all working closely together along with their reptilian allies beneath and beyond this planet. When and if the final conflict breaks out, we can expect the high-ranking 'leaders' of both sides of the Machievellian conflict to leave the so-called left-wing Communist Socialists and the so-called right-wing National Socialists to their fate, with the hope and expectation that they will slaughter each other and eliminate all resistance when they — the Roman-Bavarian 33+ Jesuit-Mason Banksters and their alien 'hosts' in Orion and Draconis as well as some collaborating factions from Sirius-B / Hale-Bopp, etc. — emerge to take control of the planet.

In reference to the Sirian collaborators, a major irony exists in the fact that the Sirius-B zealots connected with the Hale-Bopp complex — who are so determined to stage a mass landing on earth — may be largely motivated by the fact that several of their allied underground colonies on earth and in this system have been and are under attack by the Reptiloid-Grey collectivists. Apparently they believe that by supporting their secret-society allies here on earth — who are in turn intent on establishing a New World Order — they will be in a better position to defend themselves and their ancient bases and colonies here from the Draconians-Greys.

Yet, they are in fact serving what they have been led to believe are "Ascended Masters" within the Hale-Bopp complex itself, unaware that their beloved "Ashtar" hierarchy has long since been infiltrated by Orionite Dracos and Greys. In a similar manner, the Orion-backed Jesuit Lodge had managed to infiltrate the Sirian backed Masonic Lodge on earth via the Jesuit Scottish Rite, and just as the Sirians have been duped into submitting to the agenda of the Orionites in Hale-Boop, the Masons have been duped into forming an alliance with the Jesuits in the form of the Bildeberg society. If the Masonic Lodge and our own Constitutional U.S. Government can be infiltrated by Draconian-Orionite interventionists, then are those from Sirius-B who live in an even more collectivist system any more exempt from the same threat? Maybe one day they, AND WE, will learn that the Draconians are playing for keeps and that there is NO level of deception to which they will NOT sink in order to get their way. Just like us, the Sirians and their Masonic representatives on earth have been so concerned about defending themselves from an 'enemy without', that they AND WE have ignored the infiltration of the 'enemy within'. – Branton)

#QUIENES SON LOS INSEKTOS KURS Y LOS ANNUNAKIS?

But at what? The answer is at all of the caverns which, occupied by the enemy of World War II (namely, the NAZIS – Branton), are awaiting the time to deliver THEIRnuclear missiles from sanctuaries beneath the ANTARCTIC, and from cavern strongholds beneath SOUTH, CENTRALand NORTH AMERICA.

(Note: The 33+ themselves are neither against the 'right' nor 'left' wing forces, but rather control the leadership of each, so as to set the two forces against each other in a Machievellian or Hegalian scenario —

with more than a little help from their reptiloid and grey alien advisors. This plan seems to have been traced back to the MASONIC 'Pontiff' Albert Pike [who called himself 'the vice-regent of Lucifer' on earth] and his JESUIT 'deputy' Guisseppi Mazzini.

Maybe one day they, AND WE, will learn that the Draconians are playing for keeps and that there is NO level of deception to which they will NOT sink in order to get their way. Just like us, the Sirians and their Masonic representatives on earth have been so concerned about defending themselves from an 'enemy without', that they AND WE have ignored the infiltration of the 'enemy within'. – Branton)

This explains why the U.S. and Russia have such a large surplus of atomic bombs. Caverns tend to be nuclear proof, except for direct hits which mean that at least one bomb is needed per cavern, and perhaps several just in case the first one fails to make it. All those bombs going off will have a negative environmental impact on all life on the surface; but the 33+ plan to be safe and snug in their holes. Both sides could fire all their arsenals at once; but this is unlikely to happen. World War II will bePROTRACTED [like a chess-game] and both sides will agree to a standard set of rules for war. As the war progresses and the world's standard of living drops, squabbles over remaining resources will become frequent, and pointed.

States will fight states, counties will fight counties, towns will fight towns, all of which will reflect the political biases and inclinations of controlling caverns beneath. "Did you know that only group organisms such as ant hills and termite colonies — and Masonic controlled men — indulge in mass warfare? Nothing else in nature does." All group organisms, such as bee hives, use sex odors [via the queen bee] to induce conformity in the hive or colony. [Scottish Rite] Masonry is most similar to the termite colony in that both chew away at the foundations of civilization and neither can stand the light of day. Masonry may not use sex odors to induce conformity and absorption into the group organism but it does use mesmerizing, hypnotic rays that may have sexual content to it. Selfless devotion to service, faceless anonymity, slavish devotion to a noble ideal... for the good of the whole, work without compensation, profitless causes, these are the value-philosophical ideals of an and in an ant hill, a termite in a termite colony...and a mason in a Masonic organism.

Interesting is how cleverly encyclopedias talk about springs but never caverns, and that De Sota was more curious about caverns than springs. The tunnels I recently learned of that lead off from basement rooms in the old KNIGHTS OF PYTHIAS TEMPLE in Springfield, Missouri, which is in the heart of the Ozarks, gives pause for a lot of wonder and conjecture. Until recently, I thought only a few surface dwellers knew and had access to the Underworld, but it now appears to be common knowledge among those of a specific segment of the population... It's just that those who talk don't live long.

More of what the 'Aryan elite' plan for America and the world was revealed in a lecture given by the late Phil Schneider, who was murdered [strangled to death] by persons unknown shortly after giving this speech: "I love the country I am living in, more than I love my life, but I would not be standing before you now, risking my life, if I did not believe it was so. The first part of this talk is going to concern deep underground military bases and the black budget. The Black Budget is a secretive budget that garners 25% of the gross national product of the United States. The Black Budget currently consumes $1.25 trillion per [2] years. At least this amount is used in black programs, like those concerned with deep underground military bases. Presently, there are 129 deep underground military bases in the United States.

They have been building these 129 bases day and night, unceasingly, since the early 1940's. Some of them were built even earlier than that. These bases are basically large cities underground connected by high-speed magneto-leviton trains that have speeds up to Mach 2. Several books have been written about this activity. Al Bielek has my only copy of one of them. Richard Souder, a Ph.D architect, has risked his life by talking about this. He wor irtland AFB Col. Richard Doty seems to have been torn between two intelligence agendas, explaining his seemingly schizophrenic reversals in policy regarding Dulce and related matters. Some segments of U.S. intelligence wanted to declare war on the Greys and develop SDI weapons that could be used against them in space and underground, whereas others — apparently motivated by more sinister motives — desired to continue negotiations. Two years following the Dulce wars, AQUARIUS andMAJI re-established negotiations with the Greys at Dulce for the purpose it would seem of gaining continued access to mind control technology for their New World Order agenda. – Brandon) Anyway, I got shot in the chest with one of their weapons, which was a box on their body, that blew a hole in me and gave me a nasty dose of Area 51 cobalt radiation.

I have had cancer because of that. "I didn't get really interested in UFO technology until I started work at Area 51, north of Las Vegas. After about two years recuperating after the 1979 incident, I went back to work for Morrison and Knudson, EG&G and other companies. At Area 51, they were testing all kinds of peculiar spacecraft. How many people here are familiar with Bob Lazar's story? He was a physicist working at Area 51 trying to decipher the propulsion factor in some of these alien craft. Now, I am very worried about the activity of the federal government. They have lied to the public, stonewalled senators, and have refused to tell the truth in regard to alien matters.

#26:04-32:22 min.mec.AHK PART 25 THE BEST FOR THE LAST...HEHE!!!
#VLASH ASHAMEL DEMONS...INSEKTS HUMANS EPSYLON ANGELS
#YA ESTA CLARO!!!...LOS CREADORES DE LO.S. .S.E.R.E.S HUMANOS SEGUN EL LIBRO ARB LOS KURS K HAN VUELTO AHORA SON LOS INSEKTOS LOS

FUNDADORES DE LA RAZA HUMANA SEGUN MIS INVESTIGACIONES Y CON LOS K ESTOY COMUNICADO DESDE 1999 TRAS MI PRIMERA CONEXION CON LOS KURS EN EL DESIERTO DE OUARZAZAT...BASE JEDI KURS...HH LA CLAVE LA HA ENCONTRADO DANTE SANTORI EN LA PELICULA EL JUEGO DE ENDER CUANDO APARECE LA REINA INSEKTO IGUAL K LA FOTO DE LOS KURS DEL LIBRO ARB...HH 36:52 m mec VÍDEO 25 ACUERDATE DE LAS PALABRAS "ESTE ES TU FIN!" FRASE VLASH Y TODAS LAS ESTRELLAS DEL CIELO CAYERON Y APARECIO ESA CARA DE ESTRELLAS CON ESA FRASE Y LUEGO ENSEGUIDA APARECIO MI "ALIADO" KURS-INSEKTO YA PARA SIEMPRE...9 MESES DESPUES NACIO SOPHIA OPERCIONES GENETICAS ALIEN KURS...HH.
#EPSYLON UNIDOS CON LOS ARCTURIANOS Y LOS MAYA GALACTICOS...HH.
#LA UNION ENTRE LOS INSEKTOS-KURS Y LOS HUMANOS DE EPSYLON LO ENCONTRAMOS EN LA TIKRAZIA INSEKTO ...LA GEOLOGIA NEURAL Y EL SIGLO XXIX...HH.
#TODO ESTO PASARLO A "NEXUS".,..HH.

300 U.S. citizens were made ready for the 1917 Russian Revolution.
Taught the Russian language and leftwing socialist ideology, they were shipped off to Russia to form the first politburo. Plainly, this means that socialist Russia is a tool and puppet state of the United States
(Or, more exactly the Masonic 'Banksters' operating within the U.S. — as for instance the Rockefellers, who played a major role in grooming the agents of the Communist-Socialist revolution in Russia AND the agents of the National-Socialist revolution in Germany. Whether it is left-handed Socialism or right-handed Socialism — Socialism either way you look at it isTOTALITARIANISM! – Branton)

300 U.S. citizens were made ready for the 1917 Russian Revolution.
Taught the Russian language and leftwing socialist ideology, they were shipped off to Russia to form the first politburo. Plainly, this means that socialist Russia is a tool and puppet state of the United States
(Or, more exactly the Masonic 'Banksters' operating within the U.S. — as for instance the Rockefellers, who played a major role in grooming the agents of the Communist-Socialist revolution in Russia AND the agents of the National-Socialist revolution in Germany. Whether it is left-handed Socialism or right-handed Socialism — Socialism either way you look at it isTOTALITARIANISM! – Branton)

Back in 1954, under the Eisenhower administration, the 'federal' government decided to circumvent the Constitution of the United States and form a treaty with alien entities. It was called the 1954 Greada Treaty (Eisenhower administration — established contact-landings at Holloman AFB, New Mexico; and Muroc-Edwards AFB, California in 1954. This was a year after the 'Greys' had established geosynchronous orbits around our planet within two 'planetoids' that had been engineered to serve as operational bases for later abduction, implantation, cattle mutilation, base construction, and infiltration operations. – Branton), which basically made the agreement that the aliens involved could take a few cows and test their implanting techniques on a few human beings, but that they had to give details about the people involved.
Slowly, the aliens altered the bargain until they decided they wouldn't abide by it at all. Back in 1979, this was the reality, and the fire-fight at Dulce occurred quite by accident. I was involved in building anADDITION to the deep underground military base at Dulce, which is probably the deepest base. It goes down seven levels and over 2.5 miles deep. At that particular time, we had drilled four distinct holes in the desert, and we were going to link them together and blow out large sections at a time.
My job was to go down the holes and check the rock samples, and recommend the explosive to deal with the particular rock. As I was headed down there, we found ourselves amidst a large cavern that was full of outer-space (or "inner-space"? – Branton) aliens, otherwise known as large Greys. I shot two of them. At that time, there were 30 people down there. About 40 more came down after this started, and all of them got killed. We had surprised a whole underground base of existing aliens. Later, we found out that they had been living in our planet for a long

time… This could explain a lot of what is behind the theory of ancient astronauts. (Note: This report seems to reveal a limited 'perspective' on the overall 'Dulce wars' based on the experience of one man.

It appears how
#HUMAN SACRIFICES...AZTEC TIMES...2001...WE ARE NOT AT THE PEAK OF THE FOOD PIRAMID...THE FOOD CHAIN..TO PUT AN ALIEN AT THE PEAK....HH AND A HUMAN TALLER THAN THIS ALIENS...HH.

So the 33+ degrees of Masonry are the alien-interactive levels, and 'we' are meant to believe that there are ONLY 33 degrees and no more. The Scottish Rite's infiltration of the Masonic lodges challenged the domination of the more Judeo-Christian YORK Rite. The Scottish Rite can be traced back to the Jesuit college of Clermont in France, and at the core it advocates a global government and the destruction of all national boundaries, sovereignties and cultures; the dissolution of all traditional "family" structures making all children the wards of the world state; and the destruction of the idea that man has a soul — or rather that humans are merely evolved animals having no spiritual nature and therefore no need for God. In other words a homogenized collective society which does not tolerate individual expression but instead enforces absolute conformity to the controlling establishment, kind of like the system which the Greys themselves live under.

Oráculo del Iching dice:

17. Sui

Mientras continúa trabajando bajo la dirección de sus maestros permanece abierto a los cambios que le sugiere la vida. El inferior, permanece en la sombra proyectada por su superior. Las posiciones de autoridad pueden llegar a conmocionar al débil y conducirlo a su caída. El hombre sabio busca más allá de las clases, de la riqueza, de sus relaciones, de la belleza y sus devotos asociados lo son por sus auténticos valores.

Este hexagrama involucra en primer lugar en el seguimiento de las enseñanzas de un maestro en el área de interés elegida. El verdadero maestro no necesita de discípulos, simplemente enseña a quién quiera oír, permitiendo al estudiante llegar, a su vez, a convertirse en maestro. Cuídate de aquellos que alaban inmerecidamente, en búsqueda de fáciles ganancias. El débil, crece dependiente de su reaseguro y puede caer bajo su control. El éxito, llega sólo a aquellos que tienen aptitudes. La prosperidad es algo lograble. Una riqueza estable requiere de una cuidadosa programación. Antes que nada, programa tus pasos. Que tu programa tenga la necesaria flexibilidad para responder a inesperados cambios. Sé extremadamente cauteloso. Los que de ti dependen desean autoridad. Promociónalos sólo en función de sus aptitudes. El éxito seguramente vendrá. Manténte atento a equívocas concepciones de la belleza. Crece junto a tu pareja, de manera que uno no sobrepase al otro.

13:35SEPULTURA EN DIRECTO 1 HABLANDO DE ZAPATA LAMPEAO...ETC...HH.
#BAJAR MUSICA DE "STANISLAS"...HH
www.greatdreams.com/reptilian/reptile/JPEG...drWing of reptilian working with Presión Nichols.

59 53 DE "ATMOSFERA CERO" FOTO 4 DA BOOKS!!!...HH

A mega-conspiracy?

Jan Lamprecht takes quite a different view of Cook and Peary and the controversy surrounding them. Despite all the evidence against them, he believes that they were both perfectly honest, though he admits that Peary's sanctioning of a campaign of vilification against Cook is inexcusable. He believes that both reached the north pole, that their sightings of Bradley Land and Crocker Land were genuine, and that Cook's photo of Bradley Land is authentic. To salvage their reputations, however, he has to invoke a conspiracy of incredible proportions. He argues that one or more polar lands in the north polar region have in fact been discovered — not where Peary and Cook thought they were, but north of Alaska, some 5° short of the north pole — and that they lie near or just within a polar opening measuring 100 or 200 or more miles across. He claims that the military and government authorities of Russia, America, Canada, and perhaps other nations have perpetrated an unprecedented coverup to hide these revolutionary discoveries [12]!

1:05 :48 PRESTON NICHOLS INTERVIEW...HH PARA LOS LIBROS...HH. #drainpool.

0:13- 0:49-0:50-01:13 m.mec SIGUE SIGUE SPUTNIK FOTO FOR BOOKS...HUAHUAHUA!!!

#PADRE CRESPI- HH...EGIPTO QUIERE VER FUEGO!!! BABILONIA FUL!!! WASHINGTON ES LA REINA DE UR -PADRE CRESPI

I have got it all bro.

The alphabetical system was created on where our genes came from throughout the solar system which connects us to our vibration and frequency.

Our vibration and frequency is our solar magnetic fingerprint.

Frequency and vibration are invisible to the naked eye but those like me who have trained ourselves to see there's vibrations travel as frequencies before our very eyes in all directions.

Stars work as satellites relaying these vibrations all through the solar system like mobile towers. Anyway have records of all planets being lived on and as well as visited. Que le ha pasado al presidente de Grecia?Pues que fue a hablar con el resto de países y habló con LOS GRISES MAITRE y estos le enseñaron un clon de él que hablaba y se movía como el cómo advertencia para qu

e firmará la rendición de Grecia a las condiciones totalmente tiranicas y reptilianas ...ahora al ver la situacion y ver que los GRISES MAITRE Y REPTILIANOS SON LOS QUE MANDAN EN EUROPA se le ha cambiado totalmente la perspectiva y puensa k contra eso no se puede luchar...THE RESISTANCE GOES ON!!!...HH.

hacer un blog con las bases subterráneas incluidas en el libro los13dias...incluyendo la prisión militar de regañe en Argelia con sus coordenadas incluidas...muy importante...hh.

hacer un blog con las bases subterráneas incluidas en el libro los13dias...incluyendo la prisión militar de regañe en Argelia con sus coordenadas incluidas...muy importante...hh.

#I CAN KNOW IF IS OR NOT...IF READS OR NO MY MESSAGES...HH.

According to Castello, President Harry Truman was a 'High Archon' in the interplanetary lodges and one of the first U.S. Presidents to establish a secret American treaty with Greys from Alpha Draconis and Epsilon Bootes as well as with the subterranean Ashtar forces. George Bush was apparently at one point a 42nd degree Mason, according to another source who I believe to be reliable. He would have to had been considerably high in the degrees if he were involved with MJ-12 as is claimed.

24:27 de Reptilian anonymous vídeo aparece la luna y el interior mecánico...he he para los próximos libros...hh.

C76:742-56-C222/CESPEC76:732-46-CESPEC212
C75:813-127-C33 /CESPEC75:801-115-CESPEC21
C74:609-266-C89/CESPEC74:734-48-CESPEC214
C73:604-261-C84/CESPEC73:774-88-CESPEC254
C72:738-52-C218/CESPEC72:718-32-CESPEC198
C71:667-324-C147/CESPEC71:.../K110:

Another thought — just what part do the alien Greys play in the 33+ Masons' plans? Incidentally, there are reportedly several levels above and beyond the 33rd degree, mainly those which interface with and collaborate with 'alien' fraternities or secret societies below and beyond planet earth. For instance the 'Alternative 2' and 'Alternative 3' forces who, in collaboration with the Greys, have exploited and oppressed numerous 'slave worlds' throughout this sector of the galaxy.

In his writings Whitley Streiber tells of being taken during an abduction experience to another desert-like planet with ancient ruins and tall 'Grey' type beings. He encountered American military personnel on this interplanetary excursion who were dressed in military kackies, carried camcorders and other unusual equipment. These military personnel would probably have possessed a security clearance similar to one of those mentioned above.

#ALTAIR ES LA CRUZ DEL SUR!!!...HH.

Throughout this sector of the galaxy.

According to one couple who in UNICUSmagazine told of how they defected from the Alternative-3 agenda after a 'Federation' agent informed them of these facts. The Nazi 'Neu Schwabians' are deeply involved in these joint humanoid-reptiloid interstellar atrocities against the peaceful inhabitants of other colonial worlds. The atrocities of World War II were just the beginning, since the Nazi 'holocaust', if we are to believe some 'contactees', has spread beyond the surface of this planet, both within and without.

#CONSEGUIR BAJAR "RAPA NUI"....HEHE!!!.
#METER DINERO AL MOVIL ANTES DE 15 DIAS PARA MANTENER LA LINEA ...HEHE!!.
#la tradición de los magamiledonicos...hh
As for the 33-plus levels of Masonry, according to former Dulce base security officer Thomas E. Castello — who possessed one of the highest security clearances at the base, Ultra-7 — there were several security clearances above his own that the 'higher initiates' held... such asUMBRA, STELLAR, and UMS –UNIVERSAL MILITARY SERVICE.
#EL MESSENGER SALVÓ AL MUNDO...HH 7/08/2015 CAL GREG...HH
46:31 MOEBIUS REDUX..FOTO PARA LOS LIBROS...HH
cyrax and flyrax buscar...hh
.
49:15 moebius redux foto dibujo pareja...huahuahua!!!
#ya ha sido cambiado todo...vh...#ESPAÑA NO PUEDE ARROGARSE NADA YA...GRACIAS AL BLOG Y A LA TIKRAZIA ...HEMOS DADO LA VUELTA A TODO...HH.
#nos vigilan nos roban...nos encarcelan nos matan..y encima les votamos...hh.
#IR GRABANDO LOS ATAQUES SONICOS DE AQUI CON GRABADORA Y VÍDEO..HH.
They in fact DO plan to implement computer chip control akin to the much feared "mark of the beast". And indeed such a fate just MIGHT spell out ones spiritual destruction, since when/if one is totally "assimilated" they will have forever lost their power of free choice, and in so doing will have lost their "soul". The takeover has not so much been an invasion as it has been an infestation... similar to the manner of a parasite.
#"LA BALADA DE AYALA MEUX" SOBRE UN VAQUERO SUS AVENTURAS Y LIBERTAD EN LAS MONTAÑAS DE GRANADA O EN ALGUN OTRO PLANETA....HEHE!!!
#INTO NEXUS" COLOCAR REPT COMPILATION ABOUT TALON DE AQUILES"DE LOS REPTOIDES Y CARACT.FISIC.DE GRISES Y REPT ASIMISMO METER IMAGENES DEL GRIS DISECCIONADO...Y DE CONAN...SERPIENTECGIGANTE IMAGENES REPTILIANAS...HH..^^ANTES.44:19 mecHEHE!55 41!!EN FORMATO CUENTO O HISTORIA O LEYENDA NEOMEDIEVAL...HEHE!!!...
#SI TENEMOS INTERNET LO PRIMERO ES COLOCAR LOS VIDEOS DE VENEZUELA...HEHE!!!
#LA LUNA NO DA VUELTAS SOBRE SU EJE,POR ESO SOLO PODEMOS VER UNA CARA...ES EL UNICO SAT DE UN PLANETA QUE HACE ESO...LO K SE ME DECIA DE PEQUEÑO ES K LA LUNA ERA UN SAT ARTIFICIAL Y NO LLEGABA A ENTENDER ESO DEL TODO..AHORA LO VEO QUE ES TOTALMENTE ARTIFICIAL...MADRE MÍA!!!...HEHE!!!
As I understand it, maintaining control of planet earth is ESSENTIAL to the reptilian cause of conquest and control of the entire galaxy if not beyond, due to the uncommonly diverse and rare elements and resources that this planet contains, not to mention the grid-systemwhich allows for the generation of time/space gates.
For instance, one couple who was initiated into the "cult of the serpent" [Alternative-Three operations, particularly] via a friend of theirs who was a very wealthy diamond merchant, was later approached by an agent of a major space "Federation", who informed them of the many

atrocities that the "Alt-3" forces of earth were committing against the peaceful inhabitants of other worlds...

#PEACE IS HELP CHILDREN..SOPHIA ATHENEA COBOS CLIMENT..7 YEARS...HH.

who live in even deeper recesses of the earth.

Those in the underground society have access to interstellar technology and in fact our controlled society actually serves to FEED their ongoing conquests of other worlds [including the use of Montauk-like Stargate/time-tunnel technology which our planetary grid, which is unique in the universe, allows].

Some of the major population centers were deliberately established by the Masonic/hybrid elite of the Old and New "worlds" to afford easy access to already existence under ground levels, some of which are thousands of years old, and considering that Los Alamos Labs had working prototype nuclear powered thermol-bore drills that could literally melt tunnels through the earth at a rate of 8 mph 40 YEARS AGO, you can imagine how extensive these underground systems have become.

These sub-cities also offer close access to organized criminal syndicates which operate on the surface. They have developed a whole science of "borg-onomics" through which they literally nickel-and-dime us into slavery via multi-leveled taxation, inflation, sublimation, manipulation, regulation, fines, fees, licenses... and the entire debt-credit scam which is run by the Federal Reserve and Wall Street. Then there is the "under ground Wall Street" scenario itself, which I will deal with shortly.

The "dracs" not only feed upon our outward society economically, but also emotionally [emotional vampirism], psychically [implants], and even bioplasmically [blood-fests

41 22 ALEX COLLIER 2 FOTO DESPUES DE HABER ADVERTIDO SOBRE EL 11S"EN ESTE PLANETA HAY ENERGIAS MUY NEGATIVAS,REALMENTE!!!"A NICOLAS TESLA Y/O A NEXUS "PORQUE AQUI HAY ALMAS LLENAS CON PENSAMIENTOS MUY NEGATIVOS EN ESTE PLANETA".

The "dracs" have learned that the most productive slave society is one which is not fully aware of the fact that it has been enslaved, one where the false illusion of "freedom" is maintained because of the subtle and consistent stealth through which their brand of slavery is imposed upon us, analogous to the old proverbial "frog in the boiling pot" story.

their own lighting. It was very bright.

SS: Have you any information on the face on Mars?

AB: Not that I remember in the underground. There's more than one face on Mars, by the

way   . They found several.   But I remember the NASA announcements some years ago, about 2 years ago that they're receiving a low frequency radio transmission from

Mars   . It was about 50 kilohertz, if I remember correctly.   Quite a low level indicating the equipment or whatever it was that was generating the RF signal, and it was coded, was quite old and probably nearly worn out, so they were amazed there was anything still coming out of it but it was enough of a signal that they could pick it up and put it through the

computer and transcribe it. It was a warning. A warning message to humans   not to

repeat the mistakes they made.

SS: Do you have any feelings about being on Mars? What were your general impressions?

AB: We were digging in there remains of an old civilization that preceded ours and it felt very peculiar. To look at what was left at what was once a great civilization and realize that they literally died there and left everything behind and that eventually the thing shutdown. It was in the underground deliberately apparently was survival because the circle cities had long since been destroyed and they lived down there and stayed there. From what I understand of it, a number of the Martians survived whatever the attack was on the surface eventually took off for Earth and others decided to stay behind on Mars in the underground. And quite literally their progeny eventually died out and the whole race that was left behind on Mars died out. It's rather a strange feeling to realize that the remnants of a race died out in the underground totally. They just left all their hardware behind.

The Sovereign Scribe
P.O. BOX 350 McKENNA. WA. 98558
Related arti

c412f538d935
low.
Underneath most major cities, especially in the USA in fact, there exist subterranean counterpart "cities" controlled by the Masonic/Hybrid/Alien "elite". Often surface/subsurface terminals exist beneath Masonic Lodges, Police Stations, Air Ports, and Federal Buildings of major cities... and even not so "major" cities.
The population ratio is probably close to 10% of the population [the hybrid military-industrial fraternity "elite"] living below ground as opposed to the 90% living above. This does not include the full-blood reptilian species who live in even deeper recesses of the earth.
Those in the underground society have access to interstellar technology and in fact our controlled society actually serves to FEED their ongoing conquests of other worlds [including the use of Montauk-like Stargate/time-tunnel technology which our planetary grid, which is unique in the universe, allows].
Many Federations throughout the galaxy consider planet earth to be the shit-hole of the universe, the source of MOST of the imperialistic atrocities being committed against other worlds by the Alternative-Three collective [consisting of fallen astral, reptilian, and human "elite" collaborators].
Many Federations are non-interventionist, and the reptilians below know it, which is why they have taken refuge beneath major population centers, essentiallyusing us on the surface as "human shields".
Some of the major population centers were de
COMENTAR CON STEVEN ICAN ESCRIBIR UN LIBRO SOBRE ANNUNAKI..HEHE!!!
28:37 Y 43:23 DEL VIDEO DE PHIL SCHENEIDER UNDERGROUND ALIEN BASES TOTALLY IN ENGLISH TALKING ABOUT BOMBING IN 6 7 FLOOR ON WORLD TRADE CENTER..AND HE SAYS IT AT 1995!!!!...HH.
#LOS DOMOS QUE YO DIBUJABA EN 5 DE EGB SOBRE LA LUNA ERAN DOMOS REALES DONDE HABITABAN 5 MILLONES DE EXTRATERESTRES CON JARDINES PISCINAS...DE TODO COMO EN MIS DIBUJOS...CUANDO? SON 9 DOMOS CON 14KM DE ALTURA!! 36 41 ALEX COLLIER.
I would assume they are referring to the black operations based beneath Neu Schwabeland Antarctica, Dulce New Mexico, Pine Gap Australia, andMontauk NEW YORK. An interesting connection I have found is the term ULTRA... Ultra is the name of a very high security classification dealing with alien interaction, it is also the name of a secret NSA agency dealing with the same, it is also the name of the [formerly] joint operational base-network beneath the Dulce - Los Alamos area of New Mexico, it is ALSO the name of a secret Nazi team during

World War II which dealt exclusively with security for the German underground base projects in Antarctica.

#TO THE NEW BOOK: THE RESISTANCE HAS TWO LEAD GOALS :TO DESTROY MENTALLY AND SPIRITUALLY HIVE ALIEN MIND AND TO FREE TO HUMAN RACE ALL TECH ELLITE HAS AND WITH WICH ALL IS POSSIBLE...A TECH GIVEN BY THE SAME ALIENS TO GOVERNMENTS TO SLAVE US, WE,THE RESISTANCE USES TO DESTROY BOTH THEM...HH...

#NEXUS NOTAS

#IR GRABANDO LOS ATAQUES SONICOS DE AQUI CON GRABADORA Y VÍDEO..HH.

07 53 11111 SAGRADO CORAN..UN POCO ANTES.."OS HE COMPLETADO..."

#WE HAVE THE KEY...GILDA MOURA DEFINE MUY BIEN ALGUNAS RAZAS REPT Y GRISES Y LO ENLAZA CON SI SON REALIDAD O PROYECCIONES HOLOGRAFICAS ABRIENDONOS LA PUERTA PARA METER LA CLAVE DE LA 4D DE LAS ENSEÑANZAS DE ARGUELLES Y REDONDEAR EL LIBRO "NICOLAS TESLA "EL FINAL..."THE XENOMIND..."Y AL MISMO TIEMPO TODOS LOS DEMAS LIBROS DE ADELANTE HACIA ATRAS...HEHE!!!

. They control us from cradle to grave via their global economic and electronic control system, and it will continue to be so until enough of us on the surfacejoin together in an international resistance and literally invade the underground systems and begin to "kick ass and take names". I see no other way that we can once again gain our independence.

#DYATLOV PASS BASE SUBTERRANEA MILITAR RUSA DONDE EM 2014 INTERROGARON UN ALIEN EN AUDIO MAITRE.?HEHE!!!!

"NEXUS-THE BIONIC CENTAURUS" TITULO PROXIMO LIBRO..HEHE!!!

#AÑADIR AL LIBRO..."WHEN THE GRASSHOPER-MOTHERSHIP CAME CLOSE TO BASE ON EPSYLON ERIDANI..."INSEKT MOTYERSHIPS YOU UNDERSTAND?DESCRIBE THEM WITH ALL THE DETAILS...HUAHUAHUA!!!

#(S.C.S)SUPERIOR COMMAND STRUCTURE(S.C.S)..BIEN!!!

low.

Underneath most major cities, especially in the USA in fact, there exist subterranean counterpart "cities" controlled by the Masonic/Hybrid/Alien "elite". Often surface/subsurface terminals exist beneath Masonic Lodges, Police Stations, Air Ports, and Federal Buildings of major cities... and even not so "major" cities.

The population ratio is probably close to 10% of the population [the hybrid military-industrial fraternity "elite"] living below ground as opposed to the 90% living above. This does not include the full-blood reptilian species who live in even deeper recesses of the earth.

Those in the underground society have access to interstellar technology and in fact our controlled society actually serves to FEED their ongoing conquests of other worlds [including the use of Montauk-like Stargate/time-tunnel technology which our planetary grid, which is unique in the universe, allows].

Many Federations throughout the galaxy consider planet earth to be the shit-hole of the universe, the source of MOST of the imperialistic atrocities being committed against other worlds by the Alternative-Three collective [consisting of fallen astral, reptilian, and human "elite" collaborators].

Many Federations are non-interventionist, and the reptilians below know it, which is why they have taken refuge beneath major population centers, essentiallyusing us on the surface as "human shields".

Some of the major population centers were de

tion, it is also the name of a secret NSA agency dealing with the same, it is also the name of the [formerly] joint operational base-network beneath the Dulce - Los Alamos area of New Mexico, it is ALSO the name of a secret Nazi team during World War II which dealt exclusively with security for the German underground base projects in Antarctica.

As for the New York City / Wall Street "nest", during the bombing of the World Trade Center [aka World Slave Center] wherein terrorists attempted to topple one of the towers into the other, a little known fact was briefly revealed. A SIX LEVELED SUB-BASEMENT controlled by the U.S. SECRET SERVICE suffered heavy damage.

These six sub-basements, one beneath the other, may not have ended there, based on other information that I've uncovered of massive alien infestation beneath the New York City area. These sub-basements may actually serve as a MAJOR terminal between the underground society of Masonic elite, and the surface society which it controls.

There is a claim, I do not know how accurate it is, that in order to join the secret service one has to be at least a member of the 33rd degree of Scottish Rite Masonry. The Masonic connections resident within the architecture of Washington D.C., even to the phallic obelisk disrespectfully referred to as the "Washington Monument", are legendary and extensive.

For more information on the underground empire, and the Masonic and Organized Criminal connections, please refer to the following information:

The Masonic underground network: War of The Caverns

The Masonic underground bases: Interview with Alex Christopher

The Masonic underground cults: Caverns, Dungeons and Labyrinths

Respectfully yours [in the Resistance]

Branton

Branton was "saved" [born again] in 1985 and "Branton the alter ego" is apparently still involved with the underground scenarios on a nocturnal basis, trying to put together a literal "underground resistance" movement, both in the underground bases and above…

see: http://www.angelfire.com/ut/branton/redbook.htmll.

#NEXT BOOK..."ALTERNATIVE 3"LOS HUMANOS K ESTAN HACIENDO ATROCIDADES EN OTROS MUNDOS LAS HACEN CON SUS HERMANOS HUMANOS AQUI TRAIDORES Y COLABORADORES DE LOS REPTILIANOS-GRISES Y OTROS....THE RESISTANCE GOES ON!!!...HEHE!!!.

and in turn to fuel the imperialistic atrocities throughout the galaxy which have been committed with the underground military industrial [alien] complex serving as a major base of operations. .. no overt alien invasion will take place UNLESS of course a massive human resistance arises, one which might potentially place the underground networks in danger of being physically invaded by outside forces. But as it stands, the takeover is more or less complete, the draconians are already here among us and they are in control via the electronic monetary system upon which we have become totally dependent [as opposed to independent].

#BUSCAR LOS DIRIGENTES DE ALTERNATIVA 3 EN ESTOS MOMENTOS EN EL PLANETA TIERRA OCUPADO E INVADIDO Y EN EL RESTO DE PLANETAS HUMANOS Y NO HUMANOS INVADIDOS Y CONQUISTADOS CON LA AYUDA DE LAS TROPAS DEL NWO BUSCAR LOS RESPONSABLES EN TODOS LOS PAISES DEL MUNDO..GOBIERNOS..EL PRESIDENTE DE GRECIA TUVO UNA REUNION CON MAITRE ESTA SEMANA 12 DE JULIO CAL GREG 2015... DONDE SE REUNIO..BUSCAR EL LUGAR Y LOS AVISTAMIENTOS EN LA ZONA...HEHE!!!

01:28 INTO NEXUS NOTAS : LAND OF CONFUSSION LITLE DUCK OF THE RESISTANCE...HH 01 47 MUSSOLINI 01 48 AYATOLLAH...HH.

01:28 INTO NEXUS NOTAS : LAND OF CONFUSSION LITLE DUCK OF THE RESISTANCE...HH 01 47 MUSSOLINI 01 48 AYATOLLAH...HH.

from me:

I spoke with the psychic Eric Kourganis yesterday on the phone who lives presently in Miami, Florida.

I gave him the details of the cavern area.

He said that the area is quite large with many tunnels leading into other caverns. That at one time there was a society that dwelled in that region underground.

Also that the people there have blocked off entrances that lead hundreds of miles into the ground into the Hollow Earth. But it would be impossible to access these areas as the people inside would send negative energy to thwart entry.

He said that there is no danger in exploring the deeper areas, but you would make certain to have a rope/strings as becoming lost is a possibilty.
It does appear that the monks were in contact and could go deeper inside, much deeper. And that the cavern entrances to the Hollow Earth places were blocked off in the 1500's or before.
He also said there are probably many artifacts still around. So it may be a good find for you in that way.
Keep me informed on any adventures you may have.
peace
Greg

=====
Greg Gavin
Editor
Onelight.com Publishing
http://onelight.com/vrilland/index.htm
onelighta@yahoo.com
Gavin Gallery & Studio
Owner/Artist
http://jggavin.com
gallery studio: 585-298-9208
cell: 585-313-0785
183 St. Paul #301
Rochester, NY 14604
The Hollow Earth List is created by Onelight.com for the further sharing of ideas, concepts, probabilities and enlightening information from resources on the Inner Earth and Hollow Earth now available and being made available to 'surface dwellers; as well as the announcement of Onelight.com updates and coming events.
58 56 VÍDEO DE LA CLAVE EL MARXISMO MUY BUENO!!!...EL MARXISMO ES UN PENSAMIENTO QUE PROHIBE LA REBELION EL PENSAMIENTO REBELDE...HEHE!!...LA FARGUE YERNO DE MARX...ESCRIBE UN LIBRO "EL DERECHO A LA PEREZA"...HEHE!!!
#YO SOY UN ESCRITOR DE BASES SUBTERRANEAS-CAMPOS DE CONCENTRACION-GULAGS...HEHE!!!.
OFF PLANET UNDERGROUND BASES
Off Planet Underground Bases The Moon Russian bases on the Near side of the Moon: 1. 55 54′ N, 51 00′ E …East of the Sea of Cold. 2. 16 33′ N, 48 51′ E …West of the Sea of Crises. 3. 31 53′ S, 73 09′ E …Southwest of the Sea of Fertilio. 4. 72 26′ S, 67 30′ W …Near the South Pole on the SW side. 5. 50 53′ S, 57 49′ W …Near Phocylides Crater in the SW quadrant 6. 09 26′ S, 66 52′ W …Southwest of the Ocean of Stor
Southwest of the Ocean of Storms 7. 01 23′ S, 12 27′ W …Northeast of the Apollo 14 landing site. Russian bases on the Far side of the Moon: 1. 36 00′ S, 147 00′ E …Jules Verne Crater. Contains large underground lakes, plants, alien machinery, food storage, spacecraft hangers. Approximate 37,000 humans. Base is approximately the size of New York State. 2. 13 36′ S, 108 26′ E …Southwest of Pasture Crater. 3. 51 03′ N, 095 00′ E …Southwest of Compton crater. Functions: Living spaces, mining, strategic bases, particle beam research. Levels: Multiple Tunnels to: Unknown Notes: Taurus Mt. Entrance to US base Archimedes Entrance to US base North/South Poles Entrances… Read Mo.
I GREW UP AT LYRA WITH MAMOUTH'S MILK STRONG AND SMART...new booK 37 :51 min.mec. DE "UNTIL THE END OF THE WORLD" PASAR EL DISCURSO METAM-MATEMATICO DE ELLA PARA EL LIBRO DE NICOLAS...FINAL "THE XENOMIND...." MUY BUENO.

I recently received a message from a person living in
Spain. I will not give the details as I wish that he
is free to explore what he found on his farm when he
was 9.
from Gregorio:
>>>>
It is curious Gavin,
>> >>   thank you very much for your emails,
It's
>> very
>> >> nice to know we are not crazy.
>> >>
>> >>   From very young (9 years old more or
less) I
>> >> found an old cave on my farm, this farm was
>> founded
>> >> by Franciscanos (religious catholic order
at
>> >> spain), and they made stairs and roof so it
could
>> be
>> >> easily accessed.
>> >>
>> >>   Few weeks ago I found a way to get in
through
>> >> other tunnel, and I found stairs going
directly
>> >> down, I got down for at least 45 minutes,
And
>> >> wondering for the batteries I came back
up.
>> >>
>> >> I have very nice feelings when I get in,
and
>> >> wondering how could I contact its
inhabitantas I
>> >> write you these lines...
#VERY IMPORTANT...END NICOLAS...THE XENOMIND...MONTGO MOUNTAIN
NEAR XABIA (ALICANTE-SPAIN) IS A VERY ACTIVE PLACE INFRONT IBIZA
MANY ALIEN RACES...DÉNIA HAS A REPTILIAN UNDERGROUND BASE AND
BELOW THIS MOUNTAIN THERE ,S A ERIDAIEAN BASE WITH
ANDROMEDIANS...MANY REPTOIDS AND GREY RACES AS MAITRE AND OTHERS
FROM ORIONITE EMPIRE ARE IN BATLE WITH THE ERIDAIEANS TO CONTROL
THIS STRATEGICAL PLACE FOR THEIR SHIPS LANDING (EIVISSA).
B: Al Bielek
SS: Regarding your experience on Mars you walked through the time tunnel, you take a step
and you're on Mars: What did you see?
AB: Well I was not on the surface of Mars. We were in the underground. The story goes back to
the Alternative 3 book, the TV production in England outlining the fact that we have Mars
bases, one or more, Provided by a joint operation with the US. government. I do not know if the
Russians are in on it – and aliens. They are on the surface bases It's a World Government
operation really, that's not strictly the United States government. After they were on the surface
which was about 1969, they found that there where entrances to the underground sealed and
they knew there was something down there. The rumors were that there was probably artifacts
from an ancient civilization buried underground because there were a lot of remains above

ground, ruined cities that have been there byNASA's estimates maybe 300,000 years, 250,000 years. But they found the entrances all blocked, all scaled off to any underground areas.

So the word went back through communications (in the late 70's) to whomever back to the Montauk and Phoenix project, "Can you do anything about this for us? We can't get into the underground of Mars." They said, "Yes, I think we can. Give us some coordinates on the surface of the planet. We'll have to run astronomical computation." Which they did and plugged these all into the computer. They wanted two people to go and it happened to be Duncan and myself.

SS: Why two?

AB: To corroborate what the other one saw and also in case there was any problems in the underground. They didn't really know what was down there. So they sent us and we went up there in the underground. [Using the Montauk Time-Space "Tunnel" device, developed as a result of the Philadelphia Experiment. (See Scribe issues 9,13 and 14.)] There was a problem with light. We had to take lighting with us at the time. Lateron, if I remember, we found some of their light sources and turned those on. We found eventually that the last remnants of the Martians , if you wish to call them that, died in the underground between 10 and 20,000 years ago by estimate , and they left everything they had of their civilization underground. We found enormous amounts of statuary which appeared to be religious.

SS: What did they look like? How big were they?

AB: Typically 6,7,8 foot tall, stone, gems embedded in them and so forth . SS. These were of human-like people?

AB: Yes. They were quite well preserved. Then we found archives. We found a lot of scientific equipment. We found electronic equipment down there; tons and tons of stuff. And the rumor was also later that … I didn't recall until Duncan reminded me of it about a week ago. He said, "Don't forget the 17,000 metric tons of Martian gold they took out. According to his recollection of it, it was very strange gold. It was 5 times denser than ours. It was worth an unbelievable fortune. Where it went we have no idea, but it was returned to Montauk and from there it went somewhere . There were several authorized trips. And Duncan and I got the bright idea since everything was in the computer let's take a trip or two on our own and do our own exploring. So we did. After the second one it was found out and we were stopped. That was when he got into the archives and found enormous records of the civilization which was buried down there.

SS: What did you find out?
AB: He was t
>> >> write you these lines...
> Hello Greg,
>> I'm glad is interesting for you too.
>> Spain, an area with neolitic settlements at
Madrid,
>> the cave goes down more
>> than 2500 yards
>> through irregular steps, and also goes up as
It
>> crosses the sides of the
>> valley, as it should have another entrance
still
>> unfound at the upper part
>> of one side of the Valley. A old tale names
these
>> passages as 'passage of
>> the monks', as it is beleived that some monks
use to
>> used it to move around.
>> I am sure that the outter part is made by
them, but
>> as soon as the irregular
>> steps appear the aspect of the cave changes.
It is
>> funny as it looks older
>> but it will last longer... how to explain the
>> endurance and deep of the
>> feeling.
>>
>> I wonder if I would have to stay at the cave
until
>> my eyes get more used to
>> the low amount of light and turn off the
>> battery-light. But this would take
>> more than a weekend, who knows if I could do
it.
>>
Source: The Sovreign Scribe http://www.freezone.org/mc/e_conv06.htm Interview provided
courtesy of QUANTUM COMMUNICATIONS.   This is a collection of Material from the book
"Matrix III" (The Psocho-Social, Chemical, Biological, and Electronic Manipulation of Human
Consciousness), from Valdamar Valerian, First Edition Printing May 1992, Copyright 1992
Valdamar Valerian. Adress: Leading Edge Research, P.O. Box 7530, Yelm, Washington State
98597. Interview with Al Bielek 1991 Al Bielek, noted lecturer on the famous "Philadelphia
Experiment" and the time travel/mind control experiments of the "Montauk Project," recently
spoke with The Scribe interview team in Yelm. Bielek gave an update on the current use of
mind control and psychic warfare, and also offered a more detailed account of his experience in
the Montauk Project. Montauk, also known as the Phoenix Project, used Bielek and his brother
Duncan Cameron, to explore the underground cities of Mars. SS: Sovereign Scribe – AB: Al
Bielek SS: Regarding your experience on Mars you walked through the time tunnel, you take a
step and you're on Mars: What did you see? AB: Well I was not... Read More
Anyway, another incredible example of all this in the Cymatics videos is seeing almost:human-
like figures forming from the particles when certain sounds are emitted. Our bodies are also the
result of sound resonating energy into form and if our minds are powerful enough to change the

sound range of the body, it moves into another form or disappears from this dimension, altogether. This is what is called shape-shifting. It is not a miracle, it is science, the natural laws of creation. The full-blood reptilians of the lower fourth dimension can therefore make their 'human' physical: form disappear and ~ bring forward their reptilian level of existence. They shape-shift. To us in this dimension they appear human, but it's just a vibrational overcoat, Hillary Clinton appeared as a reptile, while her husband, Bill Clinton the US President, was only overshadowed, and controlled by one. This is interesting because my own research, and: that of others, has revealed Hillary Clinton to be much higher in the hierarchy than Bill, who while of a crossbreed bloodline, is a pawn in the game, to be used and discarded as necessary, It is not always that the most powerful people are placed in what appears to be the most powerful jobs.

This reminds me of yet another account from the early 1980's which was covered in a now-defunct newsletter called "THE CRYSTAL BALL", which printed a "Shaver Mystery" special edition. This edition recounted a story of Russian scientists who, searching for a meteorite impact in northern Siberia, discovered an underground facility beneath the ice and snow. Several humanoid bodies were discovered in frozen suspended animation. Some of these were revived but they proved to be far from human, but rather reptilians who had somehow shape-shifted into human form… whereas the true humans in this ancient scientific colony could not be revived. Eventually the whole scientific team was "assimilated" by the reptilians, the reptilians somehow reaching into their minds and absorbing their memories and physical features. Other scientists were called in and these in turn were assimilated/absorbed by other reptilians. The story has it that the infiltration reached even to the deepest levels of the Russian Politbureau and if true it may have paved the way for other "infiltratration" agendas involving other countries.

Between the Archuleta Mesa of New Mexico (the main headquarters of the malevolent Reptiloid forces) and Death Valley in California (below which lies the main headquarters for the benevolent humanoid forces) there are several bases where things are "out of control". These consist of huge cavern systems linked together via artificial tunnels… with main "bases" below Deep Springs, CA; Mercury, NV; Dougway SW of and Granite Mt. SE of Salt Lake City; Page, AZ; Creede and also the Denver International Airport in Colorado, etc.

The benevolent humanoid Federation forces involved are from the Andromeda and Pleiades constellations, and also Tau Ceti, Vega Lyra, Procyon, Wolf 424, Alpha Centauri, etc. The malevolent Reptiloid Empire forces are from Draconis and Orion, and also Epsilon Bootes, Zeta Reticuli II, Capella, Polaris, etc.

etween 1979 when the Dulce and Groom wars broke out leading to the take-over of "our" joint operational bases and 1989 when the reptiloids/ grays took control of the Alternative-3 bases on Luna and Mars, several of the Melchizedek bases were also attacked as the reptiloids/grays turned on these native subterranean residents. During the two-year period when the executive branch of our government broke relations with the Grays following the Dulce-Groom wars, the intelligence community split into two factions: the American-Navy backed COM-12 agency which no longer desires interaction with the Grays, but seeks to maintain contact with the Pleiadeans instead and is fighting to preserve Constitutional government; and the Bavarian-CIA backed AQUARIUS agency which seeks to maintain contact with the Grays, etc. in that they are depending on their mind control technology, abductions and implants to impose a joint human-alien fascist "New World Order" dictatorship.

. This is an important distinction. There are the 'full-bloods' who are reptilians using an apparent human form to hide their true nature, and the 'hybrids', the reptile-human crossbreed bloodlines, who are possessed by the reptilians from the fourth dimension. A third type are the reptilians who directly manifest in this dimension, but can't hold that state indefinitely. Some of the 'Men in Black' are examples of this.

Dear ;

I am an abductee who has experienced suppressed encounters with grays and humans throughout an underground system that spans the western base of the Wasatch-Rocky mountains in Utah.

Alien power plant (Photo credit: wili_hybrid)

I've had "altered state" contacts with the humans who live within the underground system, those who I refer to as the "Melchizedeks". These are members of a metaphysical lodge with connections to the deep initiatory levels of Mormonism, Masonry, the Mt. Shasta/Agharti network, Mayans, Sirius, Arcturus, Saturn, etc. They had formerly maintained cautious territorial treaties with the branch of grays/reptiloids that are native to the underground levels, and the reptiloids took advantage of this agreement in order to infiltrate their society.

Then there are the experiences of Cathy O'Brien, the mind controlled slave of the United States government for more than 25 years, which she details in her astonishing book, Trance Formation Of America, written with Mark Phillips. She was sexually abused as a child and as an adult by a stream of famous people named in her book. Among them were the US Presidents, Gerald Ford, Bill Clinton and, most appallingly, George Bush, a: major player in the Brotherhood, as my books and others have long exposed. It was Bush, a pedophile and serial killer, who regularly abused and raped Cathy's daughter, Kelly O'Brien, as a toddler before her mother's courageous exposure of these staggering events forced the authorities to remove Kelly from the mind control programme known as Project Monarch. Cathy writes in Trance Formation Of America of how George Bush was sitting in front of her in his office in Washington DC when, he opened a book at a page depicting "lizard-like aliens fro m a far off, deep space place." Bush then claimed to be an 'alien' himself and appeared, before her eyes, to transform 'like a chameleon' into a reptile. Cathy believed that some kind of hologram had been activated to achieve this and from her understanding at the time I can see why she rationalised her experience in this way. Anyone would, because the truth is too fantastic to comprehend until you see the build up of evidence. 'There's no doubt that alien~based mind programmes are part of these mind control projects and that the whole UFO-extraterrestrial scene is being massively manipulated, not least through Hollywood films designed to mould public thinking. Cathy says in her book that George Lucas, the producer of Star Wars, is an operative with NASA; and the National Security Agency, the 'parent' body of the CIA." But given the evidence presented by so many other people, I don't believe that what Bush said and Cathy saw was just a mind control programme. I think he was revealing the Biggest Secret, that a reptilian race from another dimension has been controlling the planet for thousands of years. I know other people who have seen Bush shape-shift into a reptilian.

. It seemed as if the reptilian tongue could not pronounce the word "KIN-IN-I-GIN". If a suspected reptilian infiltrator was cornered and could not bring itself to pronouncing the words, they were taken and if proven to be reptilian they were dealt with accordingly.

. It seemed as if the reptilian tongue could not pronounce the word "KIN-IN-I-GIN". If a suspected reptilian infiltrator was cornered and could not bring itself to pronouncing the words, they were taken and if proven to be reptilian they were dealt with accordingly.

The president of Mexico in the 1980s, Miguel DeLa Madrid; also used Cathy in her mind controlled state. She said he told her the legend of the Iguana and explained that lizard-like extraterrestrials had descended upon the Mayans in Mexico. The Mayan pyramids, their advanced astronomical technology and ~ the sacrifice of virgins, was inspired by lizard-like aliens, he told her."' He added that these reptilians interbred with the Mayans to produce a form of life they could inhabit. De La Madrid told Cathy that these reptile-human bloodlines could, fluctuate between a human and iguana appearance through chameleon-like abilities – "a perfect vehicle for transforming into world leaders", he said. De la Madrid claimed to have Mayan-lizard ancestry in his blood which allowed him to transform back to an iguana at will. He then changed before her eyes, as Bush had, and appeared to have a lizard-like tongue and eyes." Cathy understandably believed this to be another holographic projection, but was it really? Or was De La Madrid saying something very close to the truth? This theme of being like a chameleon is merely another term for 'shape-shifting', a theme you find throughout the ancient world and among open minded people, in the modern one too.

Contactee Maurice Doreal may add something new to all of this, with his claim that he was invited — by two "blond men" who attended one of his lectures — into an ancient neo-Mayan city under Mt. Shasta, California called Telos [interesting enough, "Telos" is also a Greek word meaning "uttermost" or "purpose"]. During later contacts Doreal was shown some ancient 'holographic' libraries beneath the Himalayas, and holographic records of a technically-advanced race of tall, blond and blue-eyed humans who ruled a vast empire where the Gobi desert now lies.

These 'Nordics' were at war with a race of reptilian or neo-saurian humanoids — velociraptor type humanoids, possibly the result of ancient genetic engineering gone out of control? — based on what at the time was the semi-tropical continent of Antarctica. The 'Nordics' literally drove the reptilian humanoids off the face of the earth, the reptiloids taking refuge in vast underground cavern systems [possibly akin to "Snakeworld", "Patala" or "Nagaloka" with it's reptilian capital "Bhoga-vita" — which is according to Hindu tradition part of a seven leveled subterranean realm stretching from Benares India to Lake Manosarowar Tibet, and inhabited by deadly reptilian humanoids called the 'Nagas']. There they developed a hive-like society in order to advanced their occult-technology.

[others have suggested that many native Americans have a specific meta-gene factor that could potentially prove to be very threatening to the reptilian agenda, which is why the "lizard people" closely monitor native Americans and attempt to keep them under their hypnotic "spell"].

01:17:00 TOOL LATERALUS TALK ABOUT 51 AREA ....TO GET TO "NEXUS"

Fortunately the Luciferian collectivists according to Revelation 12 will lose their power base in the galaxy as resistance to their atrocities increases. Unfortunately however the central command of the collective will escape to the caverns of planet earth, which will serve as their "last stand", and according to prophecy they will begin a desperate program to recruit Terrans through a European-based New World Order involving electronic mind control implants and ancient roots in the remnants of the ill-named Holy Roman Empire. All of this will be a last-ditch effort to re-gain their lost ground among the stars. In the process they will devastate much of the planet, but it will all be for naught as they will lose in the end. The question is NOT whether they will lose the war, the question is how many of US will survive these apocalyptic events. I believe that this largely depends on the collective WILL of individual human beings throughout this planet. It is not something that is set in stone.

-Branton

#LA CLAVE ESTA EN VENUS COMO SABEMOS EN 4 O 5 O 6 O 7 DIMENSION...MIRAR BARCO A VENUS LA CANCION ESCUCHARLA DE MECANO...QUE ES UN MECANO?...HH.

#RUTA DE ACCESO A NOTAS DE "NEXUS" EN EL CARPESANO CON CLIP GIGANTE DE LA PARTE DE ARRIBA SUPERIOR IZQUIERDA ENFRENTE AL ENTRAR AL HOSPICIO...CENTRO MODULAR DE LA GALAXIA EPSYLON...HH RUTA DE ACCESO...HH.

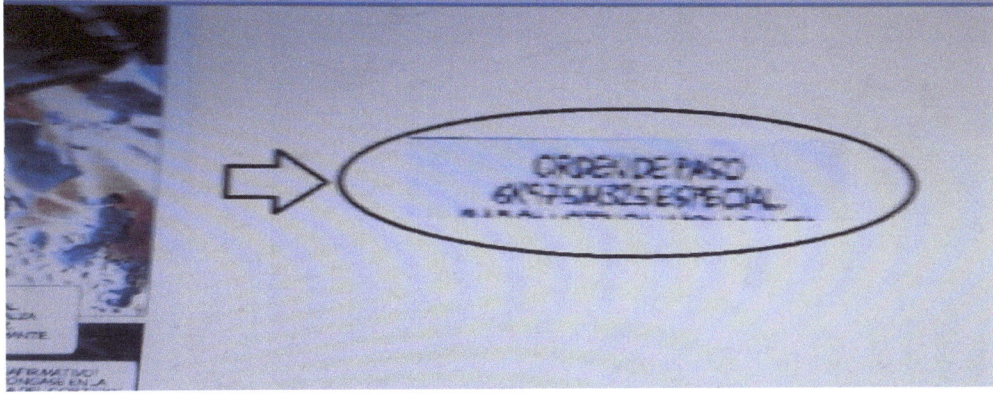

ORDEN DE PASO 6K975M325 ESPECIAL...HHH

"Bonnie, her mother (Rana Mu), her fatherRa(Mu), her sister Judy, her cousins Lorae and Matox, live and move in our society, returning frequently to TELOS for rest and recuperation. Bonnie relates that her people use boring machines to bore tunnels in the Earth. These boring machines heat the rock to incandescence, then vitrify it, thus eliminating the need for beams and supports. A tube transit tunnel is used to connect the... cities that exist in various subterranean regions in our hemisphere. The tube trains are propelled by electromagnetic impulses up to speeds of 2500 mph. One tube connects with one of their cities in the Matto Grosso jungle of Brazil. (They) have developed space travel and some flying saucers come from their subterranean bases...

"They grow food hydroponically under full-spectrum lights with their gardens attended by automatons. The food and resources of Telosare distributed in plenty to the million-and-a-half population that thrives on a no-money economy. Bonnie talks about history, of theUighers, Naga-Mayas, and Quetzals, of which she is a descendant

(Note: Many people have mistakenly identified the inhabitants of 'Telos' as being directly descended from the 'Lemurians', however Bonnie here seems to refute this by indicating that her ancestrage was other than this, possibly Meso-American and/or East-Indian? As in the case of the ancient 'antediluvian' cities of the eastern seaboard which were re-established after being abandoned by the lost 'Atlanteans', the 'Lemurians', if they existed, also seem to have been devastated in a world-wide cataclysm and their cities re-established by the Uighers, Naga-Mayas, and Quetzals and probably scattered members of other societies. As we've said earlier, the name 'Telos' is a Grecian word meaning 'uttermost', suggesting a 'possible' connection with the grecian-like Hav-musuvs of the Panamint mountains of California - Branton).

"I met Bonnie's cousin, Matox, who, like her, is a strict vegetarian and holds the same attitudes concerning the motives of government. They constantly guard against discovery or intrusion. Their advanced awareness and technology helps them remain vigilant...

"Science Fiction? Bonnie is a real person. Many have met her. Is she perpetrating a hoax? For what motive? She does not seek publicity and I have a devil of a time getting her to meetings to talk with others, but she has done so. There has been little variation in her story and her answers in the past three years. She has given me excellent technical insight on the construction of a crystal-powered generator that extracts ambient energy... Bonnie's father, the Ramu, is 300 years old and a member of the ruling council of Telos.

"Many tunnels are unsafe and closed off. All tube transit tunnels are protected and are designed to eject uninvited guests. Does Bonnie have the answers that we are looking for? I don't know... Bonnie says she would like to satisfy our need for proof and will work with me on a satisfactory answer to that problem, but she is unconcerned with whether people accept her or not. Bonnie is humorous and easy-going and well-poised, yet sometimes she becomes brooding and mysterious. She says her people are busy planning survival centers for refugees. One of these is to be near Prescott, Arizona..."

(Note: or rather below the Groom Creek area just south of Prescott, to be exact. Another 'survival center' for refugees of the world-wide cataclysms which the Telosians believe will eventually devastate the surface of the earth, is said to be below the general area of Jenny Lake, Wyoming, near the Tetons. The Tetons themselves have been the alleged home of a mysterious race, according to different sources, and extremely ancient stone 'buildings' have reportedly been found high atop these peaks - Branton).

When Bill Hamilton asked "Bonnie" to elaborate about the power- sources which her people utilize to propel the so-called "flying saucer" craft, she replied:

"...A lot of it is crystals (i.e. crystal-induced electromagnetism? - Branton), particularly the atmos

01:17:00 TOOL LATERALUS TALK ABOUT 51 AREA ....TO GET TO "NEXUS"

"Still another explorer named D.O. visited this same tunnel near Gaspar, Santa Catarina, and behind a wonderful fruit orchard saw a subterranean woman with a child in her arms reading to it aloud from a huge book written in an unknown language... After she read each sentence the child repeated the same and in this way was taught how to read. All of these subterranean cities are illuminated by strange light..."'

In relation to the apparent connection between subterranean civilizations and unidentified flying objects (Bernard and de Souza, incidentally, believed 'flying saucers' to be of subterranean origin), we will here quote from Paris Flammonde, author of 'THE AGE OF FLYING SAUCERS' (Hawthorne Books, Inc., N.Y.), who tends to confirm this hypothesis. He in turn quoted Raymond A. Palmer as a major proponent of this belief:

"...The new decade was not without a new theory, or, at least, a variation of an old one--that not only were Flying Saucers not originating from beyond the farthest reaches of our planet, they were expelled from within it...Ray Palmer wrote a lengthy article elaborating his interesting and imaginative thesis, and prefaced it with the assertion that he was prepared 'to prove that flying saucers are native to planet earth; that the governments of more than one nation (if not all of them) know this to be the fact; that a concerted effort is being made to learn all about them, and to explore their native land; and that facts already known are considered so important that they are the world's top secret...' The continuation of his contention reads:

'...is there any area on Earth which can be regarded as a possible origin for flying saucers? There are... four... the two major, in order of importance, are Antarctica andthe Arctic... the two minor areas are South America's Motto Grosso and Asia'sTibetan Highlands.'"

Raymond Bernard (actual name 'Walter Seigmeister'), writing in the Oct. 1959 issue of SEARCH Magazine, p. 48, described yet another alleged encounter with a subterranean race. What are we to make of all these stories? Are we to assume that some of the individuals who toldBernard such accounts actually made them up, as some suggest, in order to receive the 'reward' Bernard was known to offer on documentable accounts of ancient tunnels? Or, are we to accept these accounts for just what their sources claim them to be, actual encounters with a subterranean world? Bernard stated the following:

"...Last week my investigators returned and said they visited their city (i.e. the 'city' of a race of dwarf-humans whom Bernard referred to as the 'Niebelungs', who live in a subterranean region with it's own system of illumination - Branton) and are able to bring any of my American friends to visit it, but I require one condition: absolute secrecy, as I don't want governments to send armies into the tunnel to disturb these peaceful people.

"To reach them requires a 3-day journey of about 40 miles through a tunnel. This entire distance is through a tunnel carefully lined with cut stone blocks below, above and on the sides. That was quite an engineering feat. I think the tunnel was made long to keep out curiosity seekers, and only the most determined will travel that distance.

"Here is the report of my investigations: (They are two ranchers, father and son, who discovered the tunnel accidentally):

"'We left our house 5 A.M. for the tunnel on top of a mountain and reached it 3 P.M. We were tired and camped near the entrance of the tunnel. For three days we proceeded through the tunnel. We told time by our watches, as we could not tell when it was day or night. We went to sleep at 10 P.M. and awoke at 3 A.M. and continued walking. By the third day the tunnel started to go downward by steps. It was built of stone blocks on all sides. By the night of the third day the tunnel suddenly opened into a great space covered with what appeared as a sky with a yellow light that made everything luminous, like daylight. We saw a city with many houses and saw many people in the distance. They were dwarfs with long

#TÚ ERES LA NOCHE ETERNA...Y SOLO SE RECUBRIO DE CARNE DE LABIOS DE UÑAS DE PEOR SI ABRIESEMOS PERO POR TUS OJOS Y TÚ CABELLO EL TIEMPO NO HA PODIDO CUBRIR Y SALE MOSTRANDOSE ..Y SI ABRIESEMOS PIR DENTRO DE TI NO HABRIS MUSCULOS HUESOS SANGRECSINO LAVPROPIA NOCHE OSCURA Y PROFUNDA LS ETERNIDAD DEL COSMOS HH...Escribir cuento..hh.
un lenguaje mixto entre yo y branton perfect!.
BAJAR METABARONES Y VER LOS COMICS EN EL JARDIN PARA REDACTAR "NEXUS" MUY EN EL CONTEXTO DE LA VALENCIA DE 1984 CONTEXTUALUZ EL LIBRO AHI Y EN LA VALENCIA D PEREZ CASADO PERO SIN PERDERTE EN LOS DETALLES HISTORICOS SINO EN LA PSICOLOGIA DE LOS PERDONSJES MAS COMO UNA EXCUSA QUE COMO UNA INTROD. HISTORICA Y HAWAIKA...COMICS METAL HURLANT...Y ESCRITO EN BRASIL...COMO AHORA EXORZIZAR LOS MOV REPTILISNOS DE AHORS MISMAS ARMAS METAPSICOLOGICAS...AH Y LO MAS IMPORRANTE RECORDAR QUE ESTAMOS EN UN AÑO ABEJA-ARTE (GAY) Y QUE ES UN AÑO NO PARA CAMBIAR NADA SINO PARA CREAR..Y LEER OTRA VEZ "EL

GRAN PLAN DIVINO" OBLIGAO!!!... .SIN LIMITACIONES...HH..PLAZA DE OTROS ESTACION DEL NORTE MALVAROSA DISEÑADOR DE VALENCIA MONTESINOS DISCOTECA DE LA

MALVAROSA...NOMBRE...HH PERO VOLVER AQUI PARA VOLVER OTRA VEZ ALLI Y CAMBIARLO TODO LA HISTORIA...HH Y DE NUEVO VUELTA AHORA A LA GALAXIA LYRA CON XENUS....Y BRASIL 2015 EN RIO GRANDE DO SUL ESA TEXTURA...HH...METER PROTEUS EN VALENCIA...HH...EL AÑO 1981 FUE OTRO AÑO DE METAL...HH EL AÑO K VIENE SI SERA UN AÑO DE VIAJES MUCHOS VIAJES...HH.

jEXPERIENCIAS GENETICAS REPTILIANAS...NEXUS...FOTO...#geneticalreptiliansxxxc en LENOVO.Hum.

Faltas muy graves

(Entre 30.001 y 600.000 euros de sanción)Manifestaciones no comunicadas o prohibidas ante infraestructuras críticas.Fabricar, almacenar o usar armas o explosivos incumpliendo la normativa o careciendo de la autorización necesaria o excediendo los límites autorizados.Celebrar espectáculos públicos quebrantando la prohibición ordenada por la autoridad correspondiente por razones de seguridad pública.

Some people, as strange as it may sound, believe that there is a conspiracy in effect upon and beneath planet earth, one that is designed to slowly and subtly enslave us through a constant barrage of subliminal programming, economic manipulation, and preconditioning. This plan is one involving a scenario that is designed to set all countries, nations or republics against each other [eventually doing away with all sovereignties altogether and replacing them with a global religio-eco-political control system], using infiltrators who operate within the leadership ranks of all countries, especially the USA, Europe, Russia, Red China, Australia and so on. In other words our planet is the chessboard, the countries are the squares, we are the 'pawns', and the draco are the chess players. If this conspiracy continues as planned they could "divide-and-conquer" us into oblivion, with MOST of the human population being eliminated (wars, famine, plagues, etc.). THEN they plan to emerge from their underground empire consisting of multi-connected bases — Dulce, Pine Gap, Gizeh, Neu Schwabia, etc., where most of the humans have since been.

#metabarones1anexusxxxxx

#metabarones2anexusxxxxx

dos archivos de JPEG del ordenador para meterlos en el libro"NEXUS"...HH

#EL CENTRO MODULAR DE LA GALAXIA EPSYLON NO SOLO ESTÁ EN COMUNICACION CON LA CIUDAD SUBTERRANEA DE LOS INTEARERRESRES SINO DE UNAS 5 O 6 CIUDADES AQUI DEBAJO DE ESTALAGEM...HH...Y YO TEBGO LA VENTAJA DE ESTAR EN COMUNICACION CON TODAS ELLAS QUE COMO SE HACE?SUMPLEMENRE SABIENDO DE ELLAS YA ELLAS TE PONEN EN CONTACTO CON SUS MAQUINAS Y TECNOLOGIAS Y TE VISIONAN TELEVISIONES DE PLASMA GIGANTES Y TE SUGUENN PASO A PASO CON LO K HACES Y SI ESTAS HACIENDO COSAS EN INTERNET SOBRE ELLOS...PORQUE SU PRINCIPAL TRABAJO SOMOS NOSOTROS...HH...LOS QUE PRETENDEN VIVIR UNA VIDA NORMAL CONSENSUAL SON EN REALIDAD SONAMBULOS Y ZOMBIES...A MERCED DE LAS ONDAS DE TRACCION DEL REPTILUANO MAYOR EN UNA DE

ESAS CIUDADES SUB ESTALAGEM O SUS SECUACES Y AYUDANTES EN LA SUPERFICIE...HH.
til the 1950's."

Note: There are indications that some members of certain Masonic- type 'secret government' societies, such as the Rosicrucian Order, have attempted to establish contact with the subterranean residents of Mt. Shasta, although it is uncertain just what might have come of this. Several encounters with the 'Blondes' (both subterran and exterran?) have revealed 'their' own concern about what is taking place with the abductions and mutilations of human beings by the sauroid Grays, although many of these groups claim that they cannot 'interfere' with the problem due to some 'cosmic law' of non- intervention. This may be true with those 'Nordic' or 'Blonde' societies who hail from other planetary bodies, such as the Taurians, Lyrans, Eridanians, and Cetusians (the latter of whom seem to be taking the most action to help their brothers here on earth, in essence interfering with the saurian 'interferers' from the Draconis, Bootes, Reticuli, Canis, etc. constellations), and the 'Solar Tribunal' groups of Mars, Luna, Saturn, etc, and so on. However, in the case of the Telosian-Aghartian alliance, this 'non-intervention' policy would not apply, since this is their world also, and they are just as native to earth as anyone else living on this planet. In light of this fact, and especially in light of their own awareness of the reptilian-saurian threat, we would urge them (if by chance they are reading this) to reconsider such a stance and join with their fellow human brothers and sisters on the surface in defending our society from this ancient threat

Others suggest that not only in southern Utah near Page, Arizona but also in northern Utah near Dugway [Page and Dugway according to former Dulce base security officer Thomas E. Castello being the two MAJOR underground connections between the DULCE New Mexico base and the GROOM Lake Nevada base], massive societal infiltration of reptilian entities in human guise has occurred. These reports are widespread, however these "chameleoids" have been seen more profusely near Dulce New Mexico, Dugway Utah, and Area 51 Nevada. There are also rumors of draconian infiltration of the federal and state government agencies, military facilities, mental health facilities, religious organization, industrial complex, and even high levels of the police forces in the state of Utah by especially the subterraneous counterparts of the "reptilian" species who exist EN MASSE within the underground levels beneath the state, since the connections between the Dulce and Groom bases not only involve artificial tunnel passages, but also an ancient and huge natural cavern network which the "aliens" have reportedly "stolen" from native American sub-colonies in years past in remarkably similar fashion as the early "Shaverian" accounts of subterranean Tero/Dero conflicts... the final takeover of these cavern-colonies or tribes coinciding with the outbreak of the Dulce and Groom Wars which raged from between 1979 to 1985 and resulted in many of "our" joint operational multi-trillion dollar underground facilities being taken over by the Dracos, their "Grey" subordinates, and in some cases insectoid collaborators [more of our tax dollars disappearing "down the tubes" to feed these vamperial parasites]...

#EXOBIOLOGIA...EXISTE EL HOMBRE LOS ANIMALES LAS PLANTAS Y LA EXOBIOLOGIA O SERES BIOLOGICOS ALIENS AQUI EN NUESTRO PLANETA..DE HECHO LLEVAN AQUI MAS TIEMPO QUE NOSOTROS CON LO CUAL NOSOTROS SERIAMOS LOS ALIENS Y/O EXTRATERRESTRES..O UNA RAZA MAS AHORA EN PROCESO DE AUTODETERMINACION DEL RESTO DE RAZAS SIMBIOTICAS : :
#RUTA DE ACCESO A NOTAS DE "NEXUS" EN EL CARPESANO CON CLIP GIGANTE DE LA PARTE DE ARRIBA SUPERIOR IZQUIERDA ENFRENTE AL ENTRAR AL HOSPICIO...CENTRO MODULAR DE LA GALAXIA EPSYLON...HH RUTA DE ACCESO...HH.

Once we are able to access the technology that the 'elite' have STOLAN from us then we can finally break out of this 'cradle' and our energy, population, pollution, and economic problems will cease... however we should and must of course refrain from violating the sovereignty of other worlds in doing so.

#YO ENCONTRE EL SIGNIFICADO DEL UNIVERSO UNA TARDE DE VERANO HABLANDO CON JUAN ,MI ABUELO, SENTADOS SOBRE UNOS 15 SACOS DE CEBADA...MI ABUELO METIO SU MANO GRANDE Y ARRUGADA CASI NEGRA DEL

SOL PERTINAZ Y EL TRABAJO INCOLUME Y SACO UN MONTON DE DENTRO:
"AQUI ESTA EL SENTIDO DE TODO" Y DEJO QUE LOS GRANOS DE LA CEBADA
RECIEN COSECHADA VOLVIERAN AL INTETIOR DEL SACO DE TELA...COMO EL
AGUA FRESCA DE UN MNANTIAL..."Y LOS HOMBRES SOMOS LOS GRANOS DEL
SACO"SIEMPRE INTENTAMOS DESTSCAR DEL RESTO DE GRANOS PERO AL FINAL
TODOS VOLVEMOS AL INTERIOR DEL SACO SIENDO OTRS VEZ TODOS LOS
GRANOS LO MISMO" ME DIJO CON UNA AMPLIA SONRISA Y DANDO UNA SUAVE
PERO CONTUNDENTE PALMADA EN EL AIRE PARA LIMPIARSE LOS RESTOS DEL
POLVO QUE DEJABAN LOS GRANOS..M" AL FINAL NO QUEDA DE NOSOTROS NI
EL POLVO"VOLVEMOS AL ORIGEN.
Y ASI DE ESA FORMA ME EXPLICO TODO:EL ORIGEN DE LA VIDA LOS MISTETIOS
DE LA VIDA Y LA MUERTE Y LA COMPOSICION DEL UNIVERSO Y SU
DESTINO...AÑOS DESPUES PUDE CERTIFICAR TODA SU COSMOVISION ...Y QUE
COINCIDIA COMPLETAMENTE CON LAS TEORIAS MAS DESARROLLADAS DE LA
FISICA CUANTICA Y DE KA BIOTECNOLOGIA MAS ARRIESGADAS Y CON EL
SIMIL ENLAZABA A LA VEZ CON TODAS LAS "LEYENDAS" DE LOS LIBROS DEL
HINDUISMO EL "MAHABARATA"...NO EN VANO LOS APORTES DE LOS GURUS DE
LA INDIA DURANTE LA EPOCA NAZARITA EN GRANADA DEJARON HONDS
HUELLA EN AQYELLAS TIERRAS...MUY PROFUNDO QUIEN SABE POR APORTES
DE OTROS PUEBLOS ANTERIORES QUE POBLARON AQUELLAS TIERRAS Y
HABIAN ESTADO EMPARENTADAS CON LA INDIA O CON LOS POBLADORES
ORIGINALES EXTRATERRESTRES DE ESTAS Y AQUELLAS TIERRAS (INCLUIR EN
EL LIBRO "NEXUS")...HEHE!!!

Y LE PREGUNTARON AL SABIO : CUAL ES EL SENTIDO DEL UNIVERSO ..Y EL
SABIO SENTADO SOBRE UN SACO DE CEBADA EN AQUELLA TORRIDA TARDE DE
VERANO LES MIRO CON UNA SONRISA AMPLIA EN LOS LABIOS : CUANDO
COMPRENDAIS LO QUE MUESTRA LA MIRADA DE UN GATO VOLVED Y
HACEDME LA PREGUNTA ¡...TODOS SE FUERON Y AÑOS DESPUES TODOS
REGRESARON ..ALLA DOND ESTABA EL SACO DE CEBADA Y EL SABIO SENTADO
SE HABIA TRANSFORMADO EN UN CAMPO DE CEBADA QUE LLEGABA HASTA EL
MAR Y QUE DABA DE COMER A MILLONES DE PERSONAS ..ENTONCES
COMPRENDIERON Y SE EXTENDIERON POR TODOS LOS RINCONES DEL MUNDO

"..ESE SABIO ERA MI ABUELO,JUAN..A EL LE DEDICO ESTE CUENTO..THE RESISTANE GOES ON´2015!!!...HEHE!!!>

Another man, J.B., claims to be a nocturnal astral vampire-killer, part of a team of nocturnal astral/dream warriors, a type of astral special forces who have declared war on the fifth density reptilian life-force vampires — the very same fifth density dracos who have reportedly taken over powerful world leaders, having assimilated their 'hosts' into their own beings which manifest in true reptilian form as some type of "wer-drac" manifestation during their secret "blood-fests". J.B. later confirmed the claims of K.S. when he stated that reptilian infiltrators [or chameleoids – both the 5th density "snatchers" and the 3rd density "shifters"] favored a particular shopping center in Salt Lake City. Some of these fifth density snatchers may actually have gained their third density solidity by fully assimilating/absorbing their "hosts" in similar fashion as the legendary "wer" people, thus becoming third density "shifters". Actually, all three types may exist, the repti-poltergeist parasites or "SNATCHERS"; actual third density and most likely subterraneous reptiloid "SHIFTERS" – using laser holograms, technotic projection, superficial bio-phasing, etc., to accomplish the shifts; and the intermediary entities or "SNATCHER/SHIFTERS"!?

The Underworld Empire
Part 4

Regardless of "Joe's" opinion, however, there is reason to believe that influences from these nether regions can and do affect "us" in a profound way, and even the men whom Ralph and Joeencountered, whoever they were, admitted this fact.

Is there anything else which we might be able to "read into" this scenario, based on the accumulated data which we've given in previous files? The men who were encountered do confirm than an ancient (antediluvian?) race did in fact leave behind extremely sophisticated technology, and it is probably true that man in his largely unregenerate state might be influenced to destroy themselves with these sophisticated machinery if given the chance. Then again the so- called Horlocks(perhaps the same as the 'cybernized', mind-altered and controlled "Men In Black" described by John Keel and others!?) have seemingly utilized such technology without utterly destroying themselves. This could be due to the fact that their 'controllers' (the serpent races?) realize the dangers of such technology and desire to conquer without destroying that which they are conquering.

Also, man already has enough 'technology' in the form of nuclear weaponry, etc., to destroy himself many times over, but no use adding fuel to the fire as they say. As for these underground or subsurface people, they are apparently part of a race or races who discovered these recesses either hundreds or thousands of years ago, or perhaps different groups who discovered this network throughout this entire period of time. The 'horlocks' seem to be a group working under an evil influence, for instance--as we've said--possibly that of the serpent race, since there have been documented CONNECTIONS uncovered between the MIB and theSerpent Race as we have seen

I GREW UP AT LYRA WITH MAMOUTH'S MILK STRONG AND SMART...new book. "The entrance to Golondrinas is located in one of the most primitive and uncivilized areas of Mexico and local inhabitants are afraid to approach the cave because they believe it is full of 'evil spirits' which lure people to their deaths. They tell stories of people mysteriously disappearing never to be heard from again while passing near the cave entrance. These stories may be based more on fact than fiction: they are similar in some respects to UFO abduction reports. Because of its huge size, remote location, and unique geological structure, Golondrinas would be an ideal UFO base. Naturally camouflaged cBut some things that CAC refers to seem to make sense, especially the following quote in response to reports that an alien conflict resulted in a planetoid and accompanying ships crashing into Jupiter. This planetoid was reportedly arriving from a dark star outside of our system called Nemesis, in the direction of Orion — a sphere which if it had enough mass would have become our nearest stellar neighbor, yet lacking mass it condensed into a large frozen planet about the size of Jupiter. A DARK star. This planetoid was reportedly filled with 40 million Draco warriors in cryogenic freeze. The loss of this armada reportedly set back the Draco/New World Order takeover plans for planet earth considerably. It could be that this report was completely false, but even if so there is still much evidence suggesting an alien connection to the New World Order nonetheless. CAC stated:

"This Awareness indicates that this is a great setback for the New World Order, for the New World Order was a plan BY AND FOR the Reptoids, and it would have benefited greatly had the Reptoids made their invasion on earth. It would have led to the need for a New World Order…"

It makes sense. Unite all nations into a central power and you will only have to take control of a few key individuals rather than dealing with numerous stubbornly independent sovereignties. aves in other parts of the world may serve as excellent natural bases, way stations, and 'depots' for UFOs.

http://www.subterraneanbases.com/cave-and-tunnel-entrances-of-south-america/

Did the Philadelphia Experiment open up more than just a hole in hyperspace? Could it have actually opened up a doorway for other-dimensional 'Dracos' to come flooding into our world in massive numbers, even to the point of infiltrating our world and replacing key political figures in order to impose their agenda upon us? Read the following and you might just begin to wonder.

Proyectar haces de luz sobre los pilotos o conductores de medios de transporte que puedan deslumbrarles o distraer su atención y provocar accidentes.

Faltas graves

Entre 601 y 30.000 euros de multaPerturbar la seguridad ciudadana en actos públicos, espectáculos deportivos o culturales, solemnidades y oficios religiosos u otras reuniones a las que asistan numerosas personas.La perturbación grave de la seguridad ciudadana en manifestaciones frente al Congreso, el Senado y asambleas autonómicas aunque no estuvieran reunidas.Causar desórdenes en la calle u obstaculizarla con barricadas.

#LA CLAVE ESTA EN VENUS COMO SABEMOS EN 4 O 5 O 6 O 7 DIMENSION...MIRAR BARCO A VENUS LA CANCION ESCUCHARLA DE MECANO...QUE ES UN MECANO?...HH

#YO ENCONTRE EL SIGNIFICADO DEL UNIVERSO UNA TARDE DE VERANO HABLANDO CON JUAN ,MI ABUELO, SENTADOS SOBRE UNOS 15 SACOS DE CEBADA...MI ABUELO METIO SU MANO GRANDE Y ARRUGADA CASI NEGRA DEL SOL PERTINAZ Y EL TRABAJO INCOLUME Y SACO UN MONTON DE DENTRO: "AQUI ESTA EL SENTIDO DE TODO" Y DEJO QUE LOS GRANOS DE LA CENADA RECIEN COSECHADA VOLVIERAN AL INTETIOR DEL SACO DE TELA...COMO EL AGUA FRESCA DE UN MNANTIAL..."Y LOS HOMBRES SOMOS LOS GRANOS DEL SACO"SIEMPRE INTENTAMOS DESTSCAR DEL RESTO DE GRANOS PERO AL FINAL TODOS VOLVEMOS AL INTERIOR DEL SACO SIENDO OTRS VEZ TODOS LOS GRANOS LO MISMO" ME DIJO CON UNA AMPLIA SONRISA Y DANDO UNA SUAVE PERO CONTUNDENTE PALMADA EN EL AIRE PARA LIMPIARSE LOS RESTOS DEL POLVO QUE DEJABAN LOS GRANOS..M" AL FINAL NO QUEDA DE NOSOTROS NI EL POLVO"VOLVEMOS AL ORIGEN.

Y ASI DE ESA FORMA ME EXPLICO TODO:EL ORIGEN DE LA VIDA LOS MISTETIOS DE LA VIDA Y LA MUERTE Y LA COMPOSICION DEL UNIVERSO Y SU DESTINO...AÑOS DESPUES PUDE CERTIFICAR TODA SU COSMOVISION ...Y QUE COINCIDIA COMPLETAMENTE CON LAS TEORIAS MAS DESARROLLADAS DE LA FISICA CUANTICA Y DE KA BIOTECNOLOGIA MAS ARRIESGADAS Y CON EL SIMIL ENLAZABA A LA VEZ CON TODAS LAS "LEYENDAS" DE LOS LIBROS DEL HINDUISMO EL "MAHABARATA"...NO EN VANO LOS APORTES DE LOS GURUS DE LA INDIA DURANTE LA EPOCA NAZARITA EN GRANADA DEJARON HONDS HUELLA EN AQYELLAS TIERRAS...MUY PROFUNDO QUIEN SABE POR APORTES DE OTROS PUEBLOS ANTERIORES QUE POBLARON AQUELLAS TIERRAS Y HABIAN ESTADO EMPARENTADAS CON LA INDIA O CON LOS POBLADORES

ORIGINALES EXTRATERRESTRES DE ESTAS Y AQUELLAS TIERRAS (INCLUIR EN EL LIBRO "NEXUS")...HEHE!!!

#EL CENTRO MIDULAR DE LA GALAXIA EPSYLON NO SOLO ESTÁ EN COMUNICACION CON LA CIUDAD SUBTERRANEA DE LOS INTEARERRESRES SINO DE UNAS 5 O 6 CIUDADES AQUI DEBAJO DE ESTALAGEM...HH...Y YO TEBGO LA VENTAJA DE ESTAR EN COMUNICACION CON TODAS ELLAS QUE COMO SE HACE?SUMPLEMENRE SABIENDO DE ELLAS YA ELLAS TE PONEN EN CONTACTO CON SUS MAQUINAS Y TECNOLOGIAS Y TE VISIONAN TELEVISIONES DE PLASMA GIGANTES Y TE SUGUENN PASO A PASO CON LO K HACES Y SI ESTAS HACIENDO COSAS EN INTERNET SOBRE ELLOS...PORQUE SU PRINCIPAL TRABAJO SOMOS NOSOTROS...HH...LOS QUE PRETENDEN VIVIR UNA VIDA NORMAL CONSENSUAL SON EN REALIDAD SONAMBULOS Y ZOMBIES...A MERCED DE LAS ONDAS DE TRACCION DEL REPTILUANO MAYOR EN UNA DE ESAS CIUDADES SUB ESTALAGEM O SUS SECUACES Y AYUDANTES EN LA SUPERFICIE...HH.

#HOY HE SOÑADO CON EVA...IBAMOS A UNA ESCUELA EN RUZAFA Y TODO EL TIEMPO LA COGIA DE LA CINTURA...EL MEJOR SUEÑO DE MI VIDA..HUAHYAHUA!!! 2 09 2015 CAL GREG...HH

HACER UN VIDEO DEL CACTUS REPTANTE DE PATRICIA Y COLOCARLO EN VIMEO O YOUTUBE O OTRA PLATAFORMA PERO K SE VEA O UNAS FOTOS...HH IMPRESIONARA EN EL FB...HH

e - Commander X).

"Our explorer J.D. (name on file - Commander X), who is a mountain guide of the Mystery Mountain near Joinville (where there is supposed to be an entrance) said that, several times, he saw a luminous flying saucer ascend from the tunnel opening that leads to a subterranean city inside the mountain, in which he heard the beautiful choral singing of men and women, and also heard the 'canto galo' (rooster crowing), a universal symbol indicating the existence of subterranean cities in Brazil. He said that the saucer was so luminous that it lit up the night sky and converted it into daylight. On one occasion he met a group of subterranean men outside the tunnel. They were short, stocky, with reddish beards and long hair, and very muscular. When he tried to approach them, they vanished. Often he saw strange illuminations in this area at night which were probably produced by flying saucers (We use the name 'Mystery Mountain,' rather than reveal the true name of the mountain, so that unwanted outsiders will not come here to locate it). Throughout my many years of research I have accumulated a vast amount of data which would indicate that these entrances to subterranean cities abound throughout the region.

"An elderly man living in Joinville once told me that he had visited a tunnel near Concepiao in the state of Sao Paulo, and saw in the distance a marvelous subterranean city with vehicles darting back and forth, evidently traveling through tunnels from one subterranean city to another.

"Although the following report requires confirmation, it was told to me by an explorer named N.C. who said that he had visited a tunnel near Rio Casdor and had met a beautiful young woman appearing to be about 20 years of age. She spoke to him in Portuguese and SAID that she was 2,500 years old. He also met a bearded subterranean man (Note: Often humans encountered in aerial disks or subterranean caverns declare that they are extremely old by humans standards. On the surface this might sound next to impossible, unless a revolutionary scientific breakthrough on the part of these human 'aliens' has allowed them to retard the ageing process to an extreme degree, or could the possibly that they are separated from the degenerating radiations of solar rays explain their allegedly greater longevity? Another possibility would be that throughbionics/biological transplants/prosthetics, etc. the lifespan of human beings possessing advanced biological and technological sciences might theoretically be increased dramatically. Incidentally, the writer and traveller Robert Stacy-Judd in some of his booksdescribed an exploration he and others in his party made of the peripheral areas of the Loltun caves of Yucatan.

Legend says that at least one group of people, fleeing persecution, entered en masse into the massive Loltun caves and were never seen again. Stacy-Judd tells of his own encounter with a 'cave hermit' deep in the cavern chambers who claimed to be well over 1000 years old, and who said he was a guardian of the cave and of the treasures--and city?--which lay deep below in the unknown depths, 'unknown' that is, except to the strange 'hermit'. Aside from photographs of this hermit which appeared in some of his works, the author also revealed photographs of 'underground gardens' consisting of areas of the cave which contain small patches of 'jungle', watered and lit through parts of the cavern ceilings which had collapsed, exposing them to the outer world. Whether such claims of longevity are real or whether the "subterranean" people were just playing with the minds of such explorers who encountered them, is uncertain - Branton).

#SUNSHINE HAVE BLOWN MY MIND AND THE WIND BLOWING MY BRAIN...HH.
#ESTOY EN EL PAIS DE LOS ZOMBIES VAMPIROS...HUAHUAHUA!!!
#M.I.A.B.23:00MECHH.TMR.G.O.#
THE SALVATION OF THE UNIVERSE...THE RETURN OF AL ANDALUS (A.A.)...HH.
#SUNSHINE HAVE BLOWN MY MIND AND THE WIND BLOWING MY BRAIN...HH.
#ESTOY EN EL PAIS DE LOS ZOMBIES VAMPIROS...HUAHUAHUA!!!
#M.I.A.B.23:00MECHH.TMR.G.O.#
THE SALVATION OF THE UNIVERSE...THE RETURN OF AL ANDALUS (A.A.)...HH.
#SUNSHINE HAVE BLOWN MY MIND AND THE WIND BLOWING MY BRAIN...HH.
#EN LO QUE OCURRIO EN EL DESIERTO DE OUARZAZAT LAS RELIGIONES NO TIENEN NADA QUE VER...NINGUNA...ESTO QUE ES LA REALIDAD ESTÁ MAS ALLA..DE LAS RELIGIONES DE TODAS...CLARO!!!...HEHE!!!.
Y ES LA REALIDAD...ES ALGO QUE SON EMBARGO ESTA TOTALMENTE ENRAIZADO EN NUESTRA ALMA EN LO MAS PROFUNDO Y AHORA ES EL TIEMPO DE SACAR A LA LUZ ESA HISTORIA ESA LUCHA...Y TU ERES PROTAGONISTA SOPHIA TU ERES LA HIJA DEL DESIERTO...NO LO OLVIDES...HH.
#HABLAR EN "NEXUS" DE LO DEL PRIMER MINISTRO GRIEGO.
Y SI CLON...INCLUIRÑO Y LO DE EVA Y LOS EMBARAZOS DE 3 MESES....HH
Alain Villeneuve...hh.
#HABLAR EN "NEXUS" DE LO DEL PRIMER MINISTRO GRIEGO.
Y SI CLON...INCLUIRÑO Y LO DE EVA Y LOS EMBARAZOS DE 3 MESES....HH.
PONER UN VÍDEO DE YOUTUBE DE LA VALENCIA DE RICARDO PEREZ CASADO.
¡EXPERIENCIAS GENETICAS REPTILIANAS...NEXUS...FOTO...#geneticalreptiliansxxxc en LENOVO.Hum
BAJAR METABARONES Y VER LOS COMICS EN EL JATDIN PARA REDACTAR "NEXUS" MUY EN EL CONTEXTO DE LA VALENCIA DE 1984 CONTEXTUALUZ EL LIBRO AHI Y EN LA VALENCIA D PEREZ CASADO PERO SIN PERDERTE EN LOS DETALLES HISTORICOS SINO EN LA PSICOLOGIA DE LOS PERDONSJES MAS COMO UNA EXCUSA QUE COMO UNA INTROD. HISTORICA Y HAWAIKA...COMICS METAL HURLANT...Y ESCRITO EN BRASIL...COMO AHORA EXORZIZAR LOS MOV REPTILISNOS DE AHORS MISMAS ARMAS METAPSICOLOGICAS...AH Y LO MAS IMPORRANTE RECORDAR QUE ESTAMOS EN UN AÑO ABEJA-ARTE (GAY) Y QUE ES UN AÑO NO PARA CAMBIAR NADA SINO PARA CREAR..Y LEER OTRA VEZ "EL

GRAN PLAN DIVINO" OBLIGAO!!!... 😊 😊 😊 😊 .SIN LIMITACIONES...HH..PLAZA DE OTROS ESTACION DEL NORTE MALVAROSA DISEÑADOR DE VALENCIA MONTESINOS DISCOTECA DE LA MALVAROSA...NOMBRE...HH PERO VOLVER AQUI PARA VOLVER OTRA VEZ ALLI Y CAMBIARLO TODO LA HISTORIA...HH Y DE NUEVO VUELTA AHORA A LA GALAXIA LYRA CON XENUS....Y BRASIL 2015 EN RIO GRANDE DO SUL ESA TEXTURA...HH...METER PROTEUS EN VALENCIA...HH...EL AÑO 1981 FUE OTRO AÑO DE METAL...HH EL AÑO K VIENE SI SERA UN AÑO DE VIAJES MUCHOS VIAJES...HH.
#HAY MUCHAS RAZAS GRISES...UNAS SON AMISTOSAS Y OTRAS SON REPTILIANAS...HH

COLOCAR AQUI "TODAS" LAS FOTOS DEL PENDRIVE QUE TRAJO MI MADRE PORQUE FORMAN PARTE DE LA HISTORIA..LUEGO HAREMOS LA SELECCION...HEHE!!!.

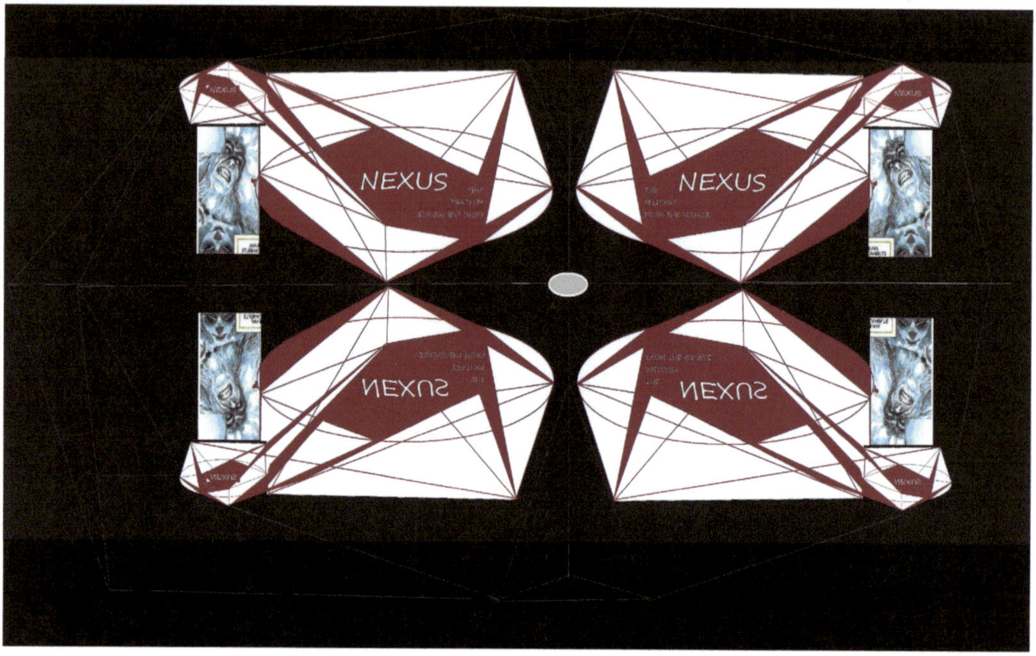

UNA NACIÓN DIVIDIDA

LA PATRIA DE DIOS:
El gobierno de coalición formado por el Cirujano General y los dirigentes de la Nueva Iniciativa Calvinista promete «limpiar la impureza» y crear un paraíso libre de humo y drogas. Se prohibirá bajo pena de muerte «la mala música, la mala comida, las malas palabras, los anticonceptivos, la pornografía y el adulterio». Los cincuenta mil Soldados Saludables y los silos de misiles nucleares del noroeste del Pacífico les proporcionan una formidable capacidad militar.

EL PAÍS DE LAS MARAVILLAS:
El parque de atracciones más grande del mundo, fue el patio de juegos de la nación. Pero ahora, los animalitos robots se han levantado en armas, exigiendo «autonomía cultural» y «el fin de la esclavitud de la inteligencia artificial». No se han confirmado los rumores sobre el asesinato y subyugación de numerosos humanos. Si el PAX atacara el País de las Maravillas, se produciría una crisis de rehenes de proporciones inconcebibles. Si se aplicara la enmienda «Sin Fronteras», morirían millones de personas.

AUTÉNTICA AMÉRICA:
Los soldados de la hamburguesería Fat Boy rodearon los extensos pastos, matando y vendiendo la carne a clientes hambrientos de buey, en flagrante violación de la enmienda 94. La empresa ilegal de comidas rápidas está bien equipada para defender su territorio: su millón de soldados y sus formidables fuerzas aéreas casi derrotaron al PAX en la Guerra del Amazonas.

TERRITORIO MEXICANO:
Aunque técnicamente forma parte de los EEUU, esta tierra superpoblada comercia abiertamente con la Auténtica América, y es la principal fuente de mano de obra barata para Fat Boy. Las fotos satélite han delatado la presencia de cientos de pastos ilegales, de manera que los analistas ya no lo llaman «la Irlanda americana», sino «el paraíso de las hamburguesas». El sentimiento antiamericano ha crecido, y los oficiales del PAX de origen mexicano se pasan al bando de Fat Boy. Una alianza entre ambas fuerzas acabaría con la soberanía de los Estados Unidos en Norteamérica.

FEDERACIÓN DE ESTADOS DE NUEVA INGLATERRA:
Militarmente débil, éste es el más pacifista de los nuevos gobiernos y el más propicio a la reunificación con EEUU. El problema es que la Federación quiere abolir prácticamente todas las enmiendas constitucionales aprobadas desde 1990. Y esos cretinos de Montaña Verde que perpetraron el saqueo de la base de datos IRS, van a tener las manos más ocupadas que nunca.

DICTADURA CAPITALISTA DE LA COSTA ESTE:
El «hombre fuerte» de Manhattan, Edward Beluga, no es una auténtica amenaza para la seguridad de EEUU. Aunque posee fuerzas convencionales, Beluga no tiene aún armas nucleares, químicas ni bacteriológicas. Además, está muy ocupado: sus soldados, agotados tras la guerra de un año entre Brooklyn y Manhattan, luchan ahora manzana a manzana contra el Movimiento Ario gay blanco racista, contra el Frente de Supremacía Negra y contra otros cincuenta movimientos separatistas.

FLORIDA:
Hace sólo dos años, el estado soleado albergaba a las comunidades más estables y mejor equipadas militarmente del país. Ahora, las líneas de suministros y comunicaciones están cortadas, y su anexión a Cuba parece inevitable.

PRIMERA CONFEDERACIÓN FEMINISTA:
La ex primera dama Amanda Nis sen, al declarar que «todo lo qu hay de malo en el mundo es culp de los hombres», se erigió en líde del poderoso Movimiento de Muje res del Sudeste, uniendo a las fac ciones enfrentadas para derroca los gobiernos del Viejo Sur. En s primer discurso, aseguró a lo EEUU que la Confederación esta blecerá relaciones diplomáticas comercio libre. Las abundante granjas del sur son una despens fundamental. Por ahora, las fronte ras permanecen cerradas para lo norteamericanos varones. Nisse prohibió la pornografía, el matrimo nio, las frases machistas y los «mo delos negativos de conducta». En l zona abundan los silos de misile pero se encuentran en mal estad En cambio, Nissen controla directa mente un cañón láser orbital, qu impide una invasión de PAX.

REPÚBLICA DE LA ESTRELLA SOLITARIA:
Texas se aferra a sus fronteras... y a sus costumbres. Con su eslogan «armas, carne y cerveza» el dentista de Dallas billy Bob Refresc estableció un gobierno basado en los «estándares de la comunidad», rechazando las enmiendas 1, 5 y 94. Su amistosa relación de codependencia con Fat Boy no deja la menor duda sobre el bando en el que militarán si llegara a estallar la guerra entre Fat Boy y PAX. Sólo el odio de Texas hacia las películas porno y las incursiones en territorio mexicano de Fat Boy impiden una alianza abierta. Texas posee un ejército de 300.000 hombres y una gran capacidad nuclear en todas las ciudades del hemisferio occidental. El próximo presidente americano tendrá que calibrar con sumo cuidado las opiniones tejanas sobre la constitución de EEUU.

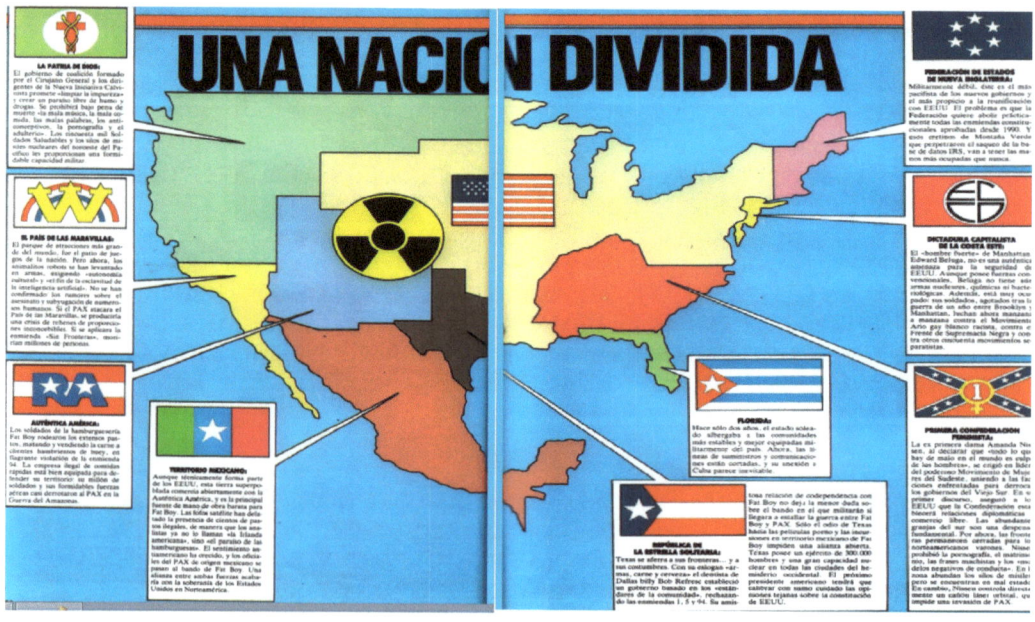

PASAR NOTAS AHK :

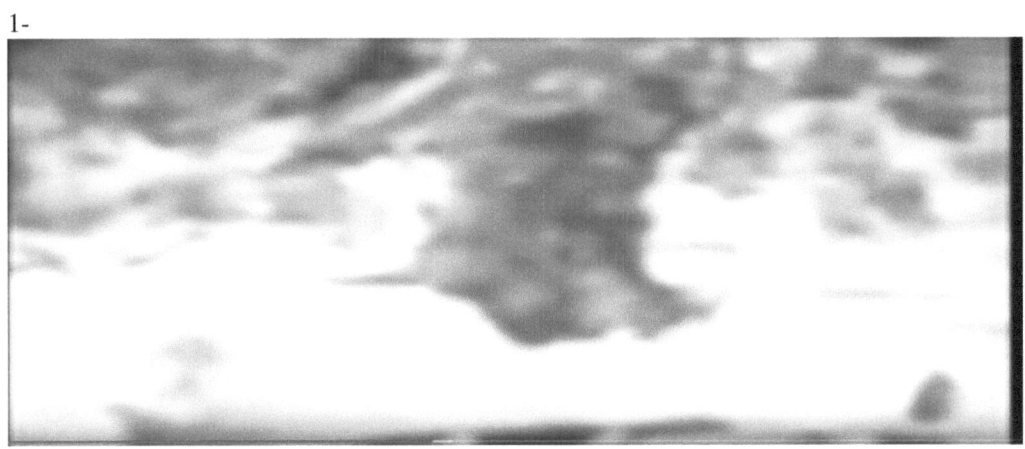

Rocks from steven the nephilin against fallen angels.
ZONA 84 NUMNERO 61 PAG 61..EL LIBRO COBRA UN NUEVO E INESPERADO
GIRO…NOS VEMOS ABOCADOS A LA ESPAÑA DE LA GUERRA CIVIL 1936-37-38-
39…HEHE!! ALLI SERÁ DONDE TENDRÁ LUGAR LA ACCION LOS ESCENARIOS
LOS PERSONAJES…PERFECTO..
LOS ERIDANIOS EN LA GUERRA CIVIL…QUE BUENA IDEA!!!...DISFRAZADOS DE
REPUBLICANOS O LAS BRIGADAS INTERNACIONALES LUCHARAN CONTRA LAS
TROPAS DE FRANCO HASTA LA MUERTE…HH..BUENO ESO YA SE INTENTO CON
"EL LABERINTO DEL FAUNO"..SI!!!...PERO ESTA VEZ ES MAS COMPLETO SIENDO
TODAS LAS HISTORIAS UNA SOLA HISTORIA…SOBRE EL CONTEXTO DE LA
GUERRA CIVIL…Y UN FINAL SORPRENDENTE….PODER RGERESAR A ESA EPOCA
Y CAMBIAR EL RESULTADO DE LA CONTIENDA…HH….Y EL NACIMIENTO DE
NEXUS XENUS EN ESPAÑA A LO PROTEUS…A TRAVESDE UNA MAQUINA
PENSANTE ESCONDIDA EN UN CASTILLO O CUEVA O SUBTERRANEO DE
MADRID…HH
1700 ANO DOMINI ROGUE RACE THE SHADOW PEOPLE 1936 SE LES VIO EN LAS
COSTAS DE ALMERIA FRAGA SE BAÑO EN ESAS MISMAS AGUAS AÑOS DESPUES
POR LOS MISMOS MOTIVOS
 EXOCOSMOBIOLOGIA EN LA GUERRA CIVIL ESPAÑOLA…ROGUE VLASH
MAITRE..RPETILIANS GRISES…HEHE!!!
EN LAS ESTRIBACIONES DE LA SIERRA DE GUADARRAMA UN GRUPO DE
MILICIANOS..HH…HEHE!!!

www.ingramcontent.com/pod-product-compliance
Lightning Source LLC
Chambersburg PA
CBHW040838180526
45159CB00001B/227